100
Science
Experiments
with **Paper**

Steven W. Moje

Sterling Publishing Co., Inc.
New York

Library of Congress Cataloging-in-Publication Data

Moje, Steven W.
    100 science experiments with paper / Steven W. Moje.
        p.   cm.
    Includes index.
    Summary: Describes how to perform 100 experiments
with paper and other materials easily found in the home, explor-
ing such topics as air, chemistry, electricity, magnetism, heat,
light, inertia, sound, and water.
    ISBN 0-8069-6391-3
    1. Science—Experiments—Juvenile literature. 2. Paper—
Experiments—Juvenile literature. [1. Paper—Experiments.
2. Science—Experiments. 3. Experiments.] I. Title.
Q164.M573   1998
507.8—dc21   98-34948
                                                        CIP
                                                        AC

10 9 8 7 6 5 4 3 2 1

Published by Sterling Publishing Company, Inc.
387 Park Avenue South, New York, N.Y. 10016
© 1998 by Steven W. Moje
Distributed in Canada by Sterling Publishing
c/o Canadian Manda Group, One Atlantic Avenue, Suite 105
Toronto, Ontario, Canada M6K 3E7
Distributed in Great Britain and Europe by Chris Lloyd
463 Ashley Road, Parkstone, Poole, Dorset, BH14 0AX, England
Distributed in Australia by Capricorn Link (Australia) Pty Ltd.
P.O. Box 6651, Baulkham Hills, Business Centre, NSW 2153, Australia
Manufactured in the United States of America
All rights reserved

Sterling ISBN 0-8069-6391-3

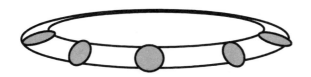

Preface

Paper is one of the most simple, versatile, available and least expensive materials known to humankind. Although most commonly used for writing, packaging, and wrapping, it is also perfect for doing science experiments. Humans have used paper or paper-like substances for thousands of years. The word *paper* comes from the word *papyrus*, a plant from which the ancient Egyptians produced a material like paper to write upon. Nowadays, wood pulp from trees is the usual source of paper fibers. To make paper, these fibers are mixed with a large amount of water. Small amounts of additives such as glue and clay are mixed in, and water is removed through wire screens. The paper fibers deposited on the screens are dried, smoothed, and cut to give many types, thicknesses, and sizes of paper.

In this book, you will learn how to do 100 exciting science experiments with paper. Experiments are organized into these categories: air, balancing, chemistry, electricity and magnetism, flying things, heat, light, motion and inertia, noise- and sound-makers, topology, water, and other experiments.

This book is easy enough for children to do, but can

be enjoyed by other people as well. The experiments in *100 Simple Science Experiments with Paper* are fun, easy, and safe to do, and can be performed using materials and equipment commonly found around the house. Teachers, parents, and children alike will delight in discovering the many ways in which paper can be used to learn and enjoy science at home and in the classroom.

Contents

Motion and Inertia Experiments

Noise- and Sound-Makers

Topology Experiments

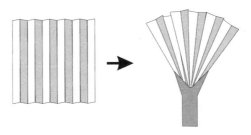

TYPES OF PAPER

There are many different sizes and weights of paper. Paper size ranges from small, 3 × 5 inch (7.5 × 13 cm) scratch pads to medium, 8½ × 11 inch (22 × 28 cm) school notebook paper, to large construction paper, artist's paper, and newsprint.* You will find these and many other sizes and kinds of paper at your local grocery, variety, arts & crafts, or hobby supply store.

Paper weight ranges from light to heavy. The heavier and thicker the paper, the stronger it is. Lightweight paper (such as onionskin or erasable typewriter paper) is good for activities where light weight is important, such as making airplanes or kites. Medium-weight paper (which includes notebook paper, scratch pads, and computer printer paper) is fairly strong and not too heavy. It can be used for most of the experiments in this book. Heavy-weight paper (for example, index cards or cardboard) is good for construction activities where strength and stability are important, such as building paper towers, and for balancing objects.

The types of paper used in this book are:

- Brown lunch and grocery bags
- Cardboard (packing boxes, cereal boxes, etc.)
- Cardboard oatmeal cartons (cylinder-shaped)

---

*These are standard sizes in the US. Paper sizes in other countries are slightly different in size. Use the closest size if these aren't available to you.

- Cardboard tubes from toilet paper or paper towels
- Computer printer paper
- Coffee filters
- Construction and drawing paper
- Crepe paper
- Dollar bill (or other paper money)
- Index cards (small and large sizes)
- Manila file folders
- Newspaper
- Notebook paper
- Note pads
- Onionskin paper
- Paper cups
- Paper plates
- Paper towels
- Tissue paper
- Toilet paper
- Waxed paper
- Writing paper

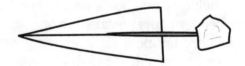

OTHER MATERIALS

Materials (in addition to paper) used for experiments in this book are:

- Bowl
- Bugs (crawling and flying)
- Buttons
- Cartons from orange juice or milk

- Coat hanger
- Coins (pennies, nickels, quarters or other coins)
- Comb for hair
- Cornstarch
- Dishwashing liquid
- Drinking straws (for soda, flexible and straight)
- Food coloring
- Fruit (soft fruit such as grape and banana)
- Glue (school glue)
- Ice cream or craft sticks
- Lemon and orange juice
- Metal washer, bolt, nut, and nail
- Paintbrush
- Paper clips (jumbo and standard sizes)
- Ping-pong ball
- Pipe cleaners
- Plastic soda bottle
- Plastic wrap
- Plates (plastic, polystyrene foam, and aluminum)
- Rocks
- Rectangular pencil eraser
- Rubber bands
- Sand, salt, sugar, rice, or other small-grained materials
- Seeds
- Soap
- Spool of thread
- String
- Thumbtack or pin
- Toothpicks
- Water and ice
- Wool

TOOLS AND EQUIPMENT

Tools and equipment used for experiments in this book are:

- Ballpoint pen
- Books
- Drawing compass
- Lamp
- Magnet (store-bought for experiment, or else use a refrigerator or shower-curtain magnet)
- Markers (water-based and permanent)
- Pencil
- Plastic dishpan
- Plastic funnel
- Plastic or glass drinking glass
- Ruler (or yardstick or meterstick)
- Scissors
- Tape (clear cellophane tape, masking tape, and duct tape)
- Toaster
- Wide-mouth jar with lid

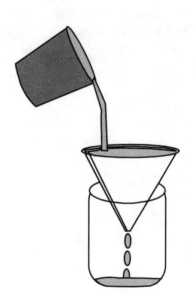

Air Experiments

PAPER FAN

You will need: Notebook paper, tape

What to do: Fold a sheet of notebook paper into an accordion shape. Pinch together the folds, 1 to 2 inches (3 to 5 cm) from one of the ends. Tape the folded end to make a handle. Grasp the handle and wave your hand back and forth. You will feel a cooling breeze on your face!

How it works: Even though you cannot see air, it is just as real as objects that are visible. Air has mass and takes up space, just like visible objects. When the paper fan moves the air, you feel the motion of the air as a breeze on your face.

More science fun: Make fans out of larger and smaller paper. Which size gives the strongest breeze? Is there a limit to the size (smallest or largest) of the fan that you can make?

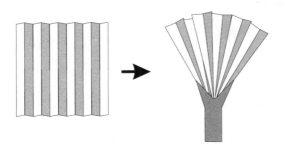

PAPER PLATE FAN

You will need: Paper plate, ice cream (craft) stick, tape

What to do: Tape an ice cream stick to the back of a paper plate. Grasp the stick and wave the plate back and forth. You will feel a cooling breeze on your face!

How it works: Just as with a paper fan, the paper plate fan pushes air when you move the fan handle. As the large round surface of the paper plate fan pushes the air, you feel it as a breeze on your face.

More science fun: Which fan (paper or paper plate) gives a stronger breeze? Make fans from larger and smaller paper plates. Which size gives the strongest breeze? Is there a limit to the size of the fan that you can make? Make fans from other types of plates (plastic or polystyrene foam).

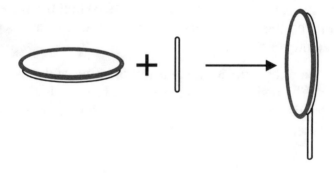

PAPER SHEET AND PAPER BALL RACE

You will need: Notebook paper

What to do: Crumple a piece of notebook paper into a ball. Drop it at the same time as you drop a flat sheet of notebook paper. The piece of paper takes longer to fall than the ball of paper.

How it works: There is more air pressing on the surface of the flat piece of paper than on the surface of the paper ball. The crumpled up ball has less air to push out of the way as it falls than the flat piece of paper does. That is why the ball hits the ground first.

More science fun: Do the experiment with different sizes and shapes of paper sheets and paper balls. Which sizes and shapes fall the fastest? Which fall the slowest?

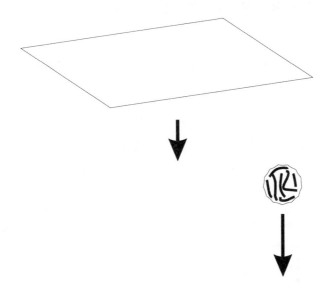

INDEX CARD FALL

You will need: Index cards

What to do: Hold up two index cards, one on edge and the other flat. Drop them both at the same time. The index card which is on edge will reach the ground first. After a short distance, the "on-edge" index card will behave as the other (flat) index card does, dropping with a random floppy motion which alternates from horizontal to vertical.

How it works: As with the flat vs. balled-up piece of paper, the object that has a greater surface area exposed to the air falls more slowly.

More science fun: Use different sizes of index cards. Which size falls the fastest? Bend an index card. Does it fall faster or slower than an unbent card?

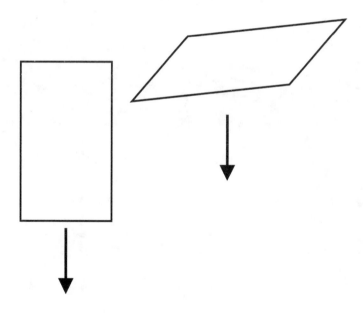

PAPER ON FALLING BOOK
OR COIN RACE

You will need: Paper, book, coin

What to do: Place a piece of paper on top of a book; make sure that the paper is smaller than the book. Drop both together. The paper will stick to the top of the book. Now drop a piece of paper by itself. It falls much more slowly. Do a similar experiment with a smaller piece of paper placed on top of a large coin (instead of on a book).

How it works: The book and coin push aside the air in front of the paper. For this reason, the paper falls at the same (fast) rate as its heavier "helper" (the book or coin). A piece of paper falling by itself cannot as easily overcome air resistance. Consequently, it falls more slowly.

More science fun: Drop different sizes of paper on different weights of books and coins. Which combinations of sizes and weights fall the fastest?

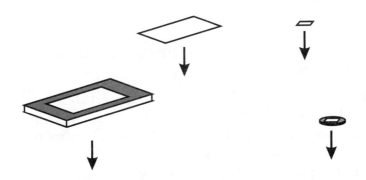

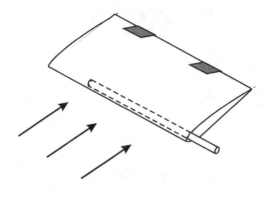

AIRFOIL

You will need: Notebook paper, straw, tape

What to do: Bend a piece of notebook paper in half. Place a straw inside the paper, along the bend. Allow 1 inch (2.5 cm) of the straw to stick out of the paper. Tape the loose ends of the paper together to make an wing-like shape or *airfoil*. Blow along the top of the airfoil. The paper will lift into the air!

How it works: Moving air has less pressure than still air. The faster that air moves, the less air pressure it has. Air passing over the curved top of the airfoil has to travel further (and thus faster) than air on the flat bottom of the airfoil. Since the air passing over the bottom of the airfoil has more pressure than air passing over the top, the airfoil is pushed up.

More science fun: Blow harder on the airfoil. Does the airfoil rise more quickly? Make a larger airfoil, using a large brown grocery bag and wooden dowel. Your breath will probably not be strong enough to cause the airfoil to rise. To create a stronger wind, use a hair dryer or the blower end of a vacuum cleaner. Which works better?

PAPER STRIP AIR LIFT

You will need: 2 × 5 inch (5 × 13 cm) strip of notebook paper

What to do: Hold a 2 × 5 inch strip of paper between your thumb and forefinger. Blow along the top of the strip. The strip will rise!

How it works: Moving air has less air pressure than still air. The still air below the strip pushes with more force than the moving air above the strip, so the strip of paper rises up.

More science fun: Which size, shape, and type (weight) of paper rises most easily?

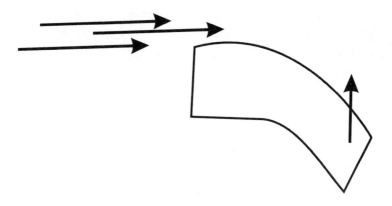

PAPER BRIDGE SAG

You will need: 5 × 8 inch (12.5 × 20 cm) index card, notebook paper

What to do: Bend down the 5 inch (12.5 cm) edges of a 5 × 8 inch index card about ½ inch (1 cm). Place the card on a tabletop, resting it on the card's bent edges. Blow underneath the card. The card will sag toward the tabletop.

How it works: The air moving underneath the card has less pressure than the still air above the card. As a result of this, the card is pressed down toward the table-top (toward the lower pressure region between the card and the tabletop).

More science fun: Do the same experiment with a lighter weight paper, such as notebook paper. How much more readily does the paper sag?

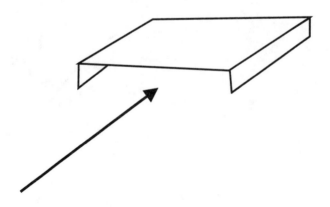

PING-PONG BALL STUCK IN FUNNEL

You will need: Plastic funnel, ping-pong ball

What to do: Place a ping-pong ball into a plastic funnel. Try to blow the ball out of the funnel. You will not be able to! The harder you blow, the more tightly the ball clings to the bottom of the funnel.

How it works: Air moving around the ping-pong ball has less pressure than the air above it, so the ball stays in the funnel.

More science fun: Use other lightweight objects, such as balls of paper or small balloons. Which remain in the funnel? Which are blown out? If any objects are blown out, do they have shapes that are less smooth than the surface of the ping-pong ball?

INDEX CARD STICKS
TO THREAD SPOOL

You will need: 3 × 5 inch (7.5 × 13 cm) index card, spool of thread, round toothpick

What to do: Push a round toothpick halfway into the center of a 3 × 5 inch index card. Place the index card on the top of a spool of thread, with the toothpick sticking into the spool's hole. Try to push away the card by blowing through the hole at the other end of the spool. You will not be able to remove it! The harder you blow, the more tightly the card sticks to the spool! Hold the spool upside down. As long as you continue to blow, the card and toothpick will stay fastened to the spool!

How it works: Air flowing around the bottom of the index card has less pressure than air above the card, so the card stays attached to the spool. (The toothpick keeps the card from slipping sideways off the spool.)

More science fun: Try other sizes of index cards. Which size works the best? Try notebook paper. When you blow through the spool, does the paper stay fastened onto the spool as easily as the index card does?

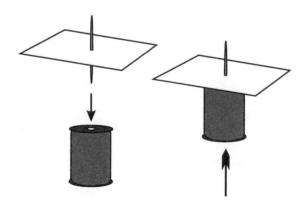

PAPER BALL AND SODA BOTTLE

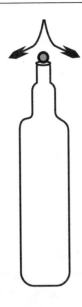

You will need: Empty plastic soda bottle, small ball of paper

What to do: Wad up a small piece of paper into a ball that is slightly smaller than the mouth of a soda bottle. Try to blow the ball into the bottle. You will not be able to!

How it works: The air in the bottle hinders the paper ball from being blown in. If the ball is just barely smaller than the opening of the bottle (and does not fall in all by itself), there is not enough space for air to escape between the surface of the ball and the inside of the mouth of the bottle, so the ball will not go in.

More science fun: Try balls made of different types of paper and of other materials, such as aluminum foil. How easily will they go into the bottle?

DRY PAPER IN A GLASS DUNKED INTO WATER

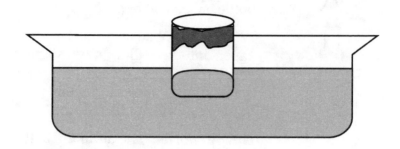

You will need: Paper, glass, dishpan of water

What to do: Wad up a piece of paper and wedge it into the bottom of a drinking glass. Turn the glass upside down and push it into a dishpan of water. The paper will not get wet!

How it works: A small amount of water is forced into the mouth of the glass when the glass is pushed below the surface of the water. However, since the air in the glass cannot escape, it presses back against the water until the pressure of the air equals the pressure of the water. The wad of paper wedged into the bottom of the glass is safely out of the reach of the water, so the paper does not get wet.

More science fun: Use taller or wider glasses. In which type of glass (taller or wider) does water rise more? Push the glass completely below the surface of the water. Does the paper still remain dry?

UPSIDE-DOWN GLASS OF WATER
DEFIES GRAVITY

You will need: Drinking glass, water, waxed paper

What to do: Fill a drinking glass with water. Carefully slide a piece of waxed paper onto the top of the glass. While holding onto the paper, quickly turn the glass upside down. Let go of the paper. The water will stay in the glass and not fall out!

How it works: Air pressure outside the glass (14.7 pounds per square inch or 2.3 kg per square centimeter) presses the waxed paper up against the mouth of the glass. If the glass is completely filled with water, the weight of the water is not enough to overcome the air pressure against the waxed paper.

More science fun: Use other types of paper, such as notebook paper or cardboard. Do any of them work as well as waxed paper? Why or why not?

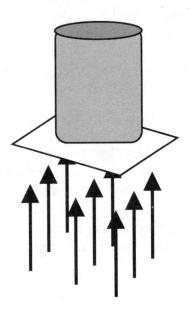

NEWSPAPER GLUES RULER
TO A TABLE

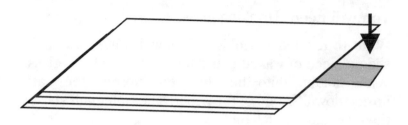

You will need: Ruler or yardstick or meter stick, newspaper, tape

What to do: Slide a ruler underneath several flat sheets of newspaper until only 2 inches (5 cm) of the ruler stick out. Strike the ruler suddenly with a thick roll of (additional) newspaper. The ruler and flat sheets of newspaper will stay in place.

How it works: Air pressure on top of the newspaper keeps the ruler in place. If the ruler is slowly pressed down, the newspaper will lift up. But if the ruler is struck hard, air cannot move out of the way fast enough and holds the ruler in place. (*Note:* for best results, use a long measuring stick such as a yardstick or meterstick, which runs the entire length of the newspaper sheets.)

More science fun: Change the length of ruler that sticks out from beneath the newspaper. What is the maximum distance that the ruler can stick out before the newspaper can no longer hold down the ruler? Put more newspapers on the first several sheets of newspaper. Which holds the ruler to the tabletop better: (1) stacking the papers straight up and down on top of each other or (2) overlapping the papers on top of each other?

WEATHER VANE

You will need: 2 straws, a jumbo size paper clip, a standard size paper clip, a 5 × 8 inch (12.5 × 20 cm) index card, scissors, plastic plate, masking tape

What to do: Make a weather vane top by cutting an arrow head and arrow tail from the index card. Tape the head and tail to the opposite ends of a drinking straw. Unbend a standard size paper clip. Stick the opened-up (pointy) end of the paper clip through the middle of the straw. For safety, cover the point with a piece of masking tape. Stick the other (round) end of the standard paper clip into a second straw.

Open up a jumbo paper clip and bend one loop at a right angle to the other loop. Tape one end of the jumbo paper clip to the bottom of an inverted plastic plate. Push the other end of the jumbo paper clip into the empty end of the second straw (see drawing). Take the weather vane outside on a breezy day. The weather vane will point in the direction from which the wind is blowing.

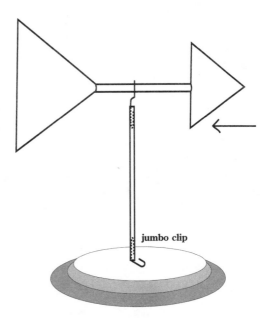

jumbo clip

How it works: The wind's force on the head and

tail of the vane causes the vane to point parallel to the direction in which the wind is blowing. Because the tail is larger than the head of the weather vane, the wind forces the tail away from it. As a consequence, the arrow head points to the direction from which the wind is coming.

More science fun: Place the weather vane in front of a window of your house on a breezy day. Does the window have air blowing in, blowing out, or neither? Do the windows on the same side of the house have the same direction of wind flow?

Balancing Experiments

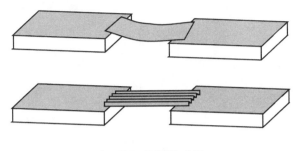

PAPER BRIDGE

You will need: 2 books, 2 pieces of notebook paper, size 8½ × 11 inches (22 × 28 cm)

What to do: Place two books a distance apart from each other that is about ¾ the length of a piece of notebook paper. Lay a piece of paper between the books. The paper will sag to the top of the table, not supporting its own weight. Fold a second piece of paper lengthwise four times into Z (accordion) shapes (see drawing). Rest the folded paper on the books as shown. The paper will remain straight, supporting its own weight. It will even support the weight of other objects, such as additional pieces of paper, paper clips, pencil, erasers.

How it works: Folded paper is resistant to bending. A piece of paper when folded can support more weight than a flat piece of paper.

More science fun: Explore different weights and lengths of paper. What size and type of paper make the best paper bridge?

PAPER TOWER

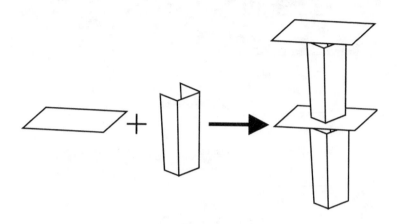

You will need: 20 to 50 sheets of notebook or scratch paper

What to do: Build a tower made from stacked pieces of paper. Fold half of the papers into V, Z, or U-shaped columns (U shapes are shown). These are the vertical (straight up and down) pieces of your paper tower. The paper sheets you didn't fold are the horizontal pieces of your tower.

How it works: The folded pieces of paper are more rigid (stiff) than are the flat pieces. Pieces of paper folded in V, Z, or U-shaped columns resist bending more than flat sheets of paper do. The flat sheets provide a stable surface for the columns to rest on.

More science fun: How high a tower can you make? Do large pieces of paper work better than small pieces of paper? How large or small can the pieces be?

PAPER BLOCKS

You will need: Index cards, tape

What to do: Fold index cards into three-dimensional shapes with triangular, square, and circular cross-sections. Tape the ends of each card together to maintain its shape.

How it works: The harder it is to bend a block, the stronger and sturdier it is. The stronger the block, the taller the structures which can be made from that shape of block. Sheets of paper bend more easily than paper blocks do because the paper fibers in flat sheets are freer to move around than the paper fibers in the blocks. You can see this effect when you crease a piece of paper. Creasing actually breaks some of the fibers in paper. These broken fibers are not able to fully return to their uncreased position. Just try to unfold a crumpled-up ball of paper and return it to its original flat state. You will not be able to!

More science fun: Which shapes make the strongest blocks? Make short and long tubes. Which are stronger: short tubes or long tubes?

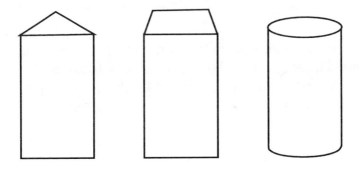

BALANCING BLOCKS

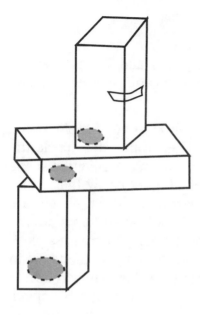

You will need: Index cards, tape, small stones

What to do: Fold index cards into three-dimensional shapes that have square or rectangular cross-sections. Tape the edges of each index card together to keep its shape. Tape small stones (or other small heavy objects, such as coins) inside the blocks. Stack the blocks. See how far off center you can stack them.

How it works: Heavy objects (stones or coins) inside the blocks change the blocks' center of gravity (center of balance). If the heavy ends are clustered around the center of the stack of blocks, some very interesting structures can be created! Make some crazy towers that look as if they're about to tip over!

More science fun: See how high and how fast you can build a tower. Place flat index cards on top of the tower. How many additional cards can be stacked on top of the balancing blocks before the whole thing tips over?

MOMMY & BABY
HAPPY FACE BALANCERS

You will need: 2 index cards, flexible straw, tape, scissors

What to do: Draw a picture of Mommy's face on an index card. Draw Baby's face on another index card. Cut out both faces. Tape one end of a straw to the bottom of Mommy's face. Tape the other end of the straw to the top of Baby's face. Balance Mommy's chin on your fingertip!

How it works: The reason that Mommy's face doesn't fall off your finger is that her face is balanced by Baby's face. The center of gravity of Mommy and Baby is below your fingertip. In order for Mommy's face to fall off your finger, the center of gravity would have to be raised. Since a center of gravity always prefers to be as low as possible, Mommy's face and Baby's face remain balanced on your fingertip.

More science fun: Draw Daddy's face (bigger than Mommy's) on an index card, cut it out, and substitute it for Mommy's face. Does Baby's face now need to be bigger in order to balance the Daddy's face?

BALANCING BIRDS

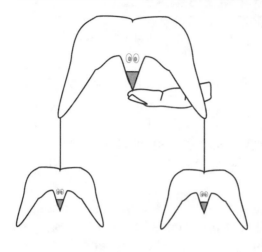

You will need: Three 5 × 8 inch (12.5 × 20 cm) index cards, tape, thread

What to do: Cut three birds from 3 index cards. Cut two equal sized pieces of thread. Tape a thread to the center back of each of 2 of the birds and tape the other end of each thread to a wing tip of the third bird (see drawing). Balance the beak of the third bird on the end of your fingertip.

How it works: In order for the center bird to fall off your fingertip, one of the two lower birds would first have to rise. This would cause the center of gravity to go up. Since the three birds want to keep their overall center of gravity as low as possible, the center bird remains on your fingertip.

More science fun: Try different sizes of birds and different lengths of thread. How many birds can you hang from the center bird? How large can they be? How small?

U-SHAPED PAPER BALANCER

You will need: Large index card, 5 × 8 inches (12.5 × 20 cm), 2 pennies (or other small coins), 4 toothpicks, tape

What to do: Cut a U shape from a large index card. Tape a toothpick to the center bottom of the U, parallel to the two sides of the U. Tape a penny to each end of the U. Stiffen the U by taping three more toothpicks to it, as shown. Turn the U upside-down. Balance the toothpick on the end of your fingertip or on a pencil eraser. Swivel the U around on your fingertip. It will turn without falling off!

How it works: The U swivels around without falling off your fingertip because its center of gravity is below the tip of the toothpick. The U does not tip over because it tends to keep its center of gravity as low as possible.

More science fun: Experiment with objects lighter than pennies, such as buttons. How lightweight can the object be before the U tips over?

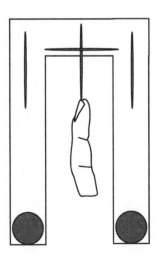

PENCIL AND
NEWSPAPER BALL BALANCER

You will need: Pencil, 2 pipe cleaners, newspaper, masking tape

What to do: Wad up a piece of newspaper into a ball and tape it to keep the ball from opening. Tape the ball to the end of a pipe cleaner. Repeat with a second piece of newspaper and pipe cleaner. Tape the other ends of each pipe cleaner 1 inch (2.5 cm) above the point of a pencil. Balance the point of the pencil on your fingertip.

How it works: The pencil balances on your fingertip because the center of gravity of the pencil, pipe cleaners, and newspaper balls is below the tip of the pencil. The pencil does not fall over because it wants to keep as low a center of gravity as possible.

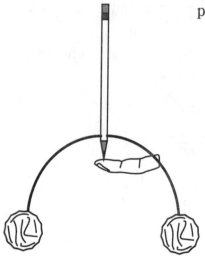

More science fun: Change the length of the pipe cleaners. Change the size and weight of the newspaper balls. Try to push the pencil over. How easy (or hard) does it become when the length and size or weights increase?

PAPER CLIP AND
PAPER BALL BALANCER

You will need: Jumbo paper clip, 2 pipe cleaners, newspaper, masking tape

What to do: Tape two pipe cleaners ½ inch (1 cm) from one end of a jumbo paper clip and on opposite sides (see drawing). Tape a newspaper ball to the free end of each of the pipe cleaners. Balance the paper clip on your finger.

How it works: The paper clip balances on your fingertip because the center of gravity of the combined objects is below the tip of the paper clip. The paper clip does not fall over because it wants to keep as low a center of gravity as possible.

More science fun: Substitute a chain of paper clips for each of the newspaper balls. How many paper clips does it take to give as much stability as the newspaper balls have?

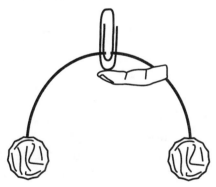

BALANCING NEWSPAPER ROLLS

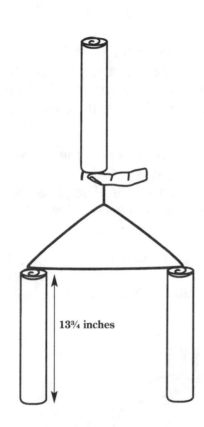

13¾ inches

You will need: Newspapers, masking tape, coat hanger

What to do: Take 4 sheets of 23 ×27.5 inch (58 × 70 cm) newspaper and fold them together in quarters so they are 11½ ×13¾ inches (29 × 35 cm). Roll them on the 13¾ inch side into a roll as seen in the diagram. Tape the roll so it doesn't unroll. Make 2 more rolls the same way. Tape two of the rolls to the corners of a coat hanger (see drawing). Insert the hook of the coat hanger into an end of the third tube. Balance the end of this tube on your fingertip.

How it works: The center newspaper roll balances on your fingertip because the center of gravity of the combined objects is below the bottom of the roll. The roll does not fall over (unless tipped excessively) because it tends to keep as low a center of gravity as possible.

More science fun: See how heavy the center roll can be before the balancing newspaper rolls structure tips over and falls off your finger.

DIVING BOARD MAN

You will need: 2 index cards, ice cream (craft) stick, flexible straw, tape

What to do: Cut out a man and a shark from two index cards. Tape the short end of a flexible straw to the man's ankles, and the other end of the straw to the mouth of the shark. Bend the straw at a 30 to 45 degree angle, so the shark is directly below the man. Place the man's feet on the end of a craft stick plank. Hold the other end of the stick in your hand. The man will not fall over, but will instead tip back and forth as he peers nervously over the edge of the board, watching the shark!

How it works: The man balances on the end of the board without falling off because the shark keeps the center of gravity below the man's feet. The man stays on the board (unless someone *pushes* him off, or he collapses and dies of fright) because he tends to keep as low a center of gravity as possible.

More science fun:
If the man is too heavy he will fall off the board. Try different sizes and weights of man and shark to see what weight of each is required for him to stay on the board!

TOOTHPICK TOP

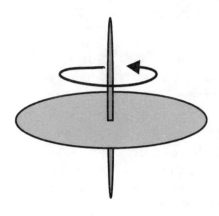

You will need: Index card, toothpick, scissors, tape.

What to do: Cut a circle from an index card. Next, push a toothpick through the center of the circle to make a top. Tape the toothpick to the circle. Grasp the toothpick and spin the top.

How it works: The top spins because of centripetal force, which stabilizes the spinning disk. When the top slows down enough, it loses its stability and falls over. You can observe this same thing with a bicycle wheel. When your bicycle is going fast enough, it stays upright. However, when it goes too slow, the bicycle falls over, since the spinning wheels can no longer counteract the force of gravity, which wants to pull the bike over onto its side to give the lowest possible center of gravity.

More science fun: Try different sizes of circles cut from index cards. How small or large can the circle become before the top no longer spins well? What size of circle gives the best spinning top?

CARDBOARD TUBE TUMBLER

You will need: Cardboard tube from toilet paper, rock (or large metal nut or bolt), index card, masking tape

What to do: Trace the circular end of the cardboard tube onto an index card in two places. Cut out the circles. Tape a circle to one end of the tube. Put a rock or large metal nut or bolt into the tube. Tape the second circle to the other end of the tube. Place the tube upright at the top of a flight of stairs. Tip it over (don't roll it). The tube will flip-flop end over end as it tumbles down the stairs.

How it works: The rock or other heavy object inside rolls from one end to the other of the tube when the tube goes down the stairs. The rock tends to stay as low to the ground as possible because of gravity. As the tube tumbles, the momentum of the rock carries it to an unstable top-heavy position; the tube corrects itself by going to a stable bottom-heavy position. As long as the cardboard tube is tumbling quickly enough, the tube finds itself in the top-heavy position again, and the end-over-end tumbling process repeats.

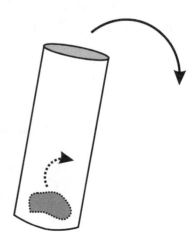

More science fun: What is the best weight and shape of rock or bolt or nut to use in the cardboard tube to give the best and longest tumbling action?

Chemistry Experiments

SOAP BOAT

You will need: Index card, toothpick, soap, duct tape or other waterproof tape

What to do: Cut a triangle from an index card. Bend (crease) the card slightly down the middle. Duct tape a toothpick to one end of the crease, so that half of the toothpick extends from the triangle. Stick a small piece of soap on the toothpick. Place the boat in a tray of water. The boat will move!

How it works: Soap breaks the surface tension of water. Water becomes "thinner" where the soap dissolves. The laws of motion say that for every action there is an equal and opposite reaction. As the soap molecules push away from the boat and into the water, the boat pushes away from the soap that has just dissolved.

More science fun: Try different types and brands of soap to see which works the best. Does soap that contains additives (such as skin conditioners) drive the boat forward as readily as a "pure" soap without additives, such as Ivory®?

PAPER YELLOWING

You will need: Different types of paper (newspaper, scratch paper, notebook paper, and good quality white printer paper)

What to do: Bend pieces of different types of paper in half. Place the pieces of paper on a bright sunny window ledge, with half of each sticking up and the other half lying flat. Check the papers every few days to see which paper discolors the most. You can tell the degree to which this occurs by comparing the part of the paper that gets the Sun's rays with the part underneath that doesn't. Low-quality paper such as newspaper will discolor (turn from white to yellow) faster than better quality paper.

How it works: Low-quality paper is made using a process that has more acid in it. Acid, when heated by sunlight, causes paper to discolor more quickly. In addition, low-quality "white" paper is usually not as white as more expensive white paper. The darker the paper, the more sunlight it absorbs.

More science fun: Cover part of each piece of paper with an object such as a toy block or paper doll. See how many days it takes for sunlight to create an outline of the object on your paper.

HIDDEN MESSAGE

INVISIBLE INK

You will need: Notebook paper; paintbrush, pipe cleaner, or toothpick; lemon juice; heat source (such as a lamp or a toaster)

What to do: With lemon juice, write a secret message on a piece of notebook paper using a paintbrush, pipe cleaner, or toothpick to write. Allow the paper to dry for a couple of hours. Then hold it over a lamp or toaster to warm it up. The message will magically appear!

How it works: After the lemon juice dries, a residue of organic solids is left behind on the paper. When heated, the solids char and darken, revealing the writing.

More science fun: Try other liquids such as orange juice, apple juice, and milk. Do they work as well as lemon juice? Also try other kinds of paper. Which types of paper work best? Does a more absorbent paper such as newspaper work as well as a less absorbent paper such as notebook paper?

Electricity and Magnetism Experiments

TISSUE PAPER DOLL'S STATIC CLING

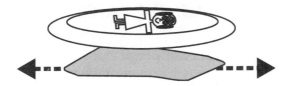

You will need: Tissue or toilet paper, scissors, piece of woolen fabric, plastic or polystyrene foam plate

What to do: Cut a small doll shape from tissue or toilet paper. Place the doll on top of an inverted plastic or polystyrene foam plate (see drawing). Turn the plate over. The doll, now on the under side of the plate, will fall off. Place the doll back on top of the inverted plate. Rub the plate from below with a piece of woolen fabric. Turn the plate over so the doll is underneath again. Now the doll will stick to the plate!

How it works: When the plate is rubbed with the woolen fabric, it develops a charge (because of static electricity). The doll is attracted to the charge and sticks to the plate.

More science fun: Put more dolls on top of the plate. Rub the plate with the woolen fabric and turn the plate over. How many dolls can you add before they fall off?

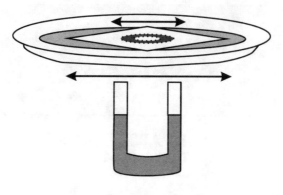

MAGNETIC PAPER

You will need: Paper or index card, metal object containing iron or steel (such as a metal washer or paper clip), tape, magnet, plastic or aluminum plate

What to do: Place the iron-containing metal object on a piece of paper or on an index card. Fold the paper over the metal piece. Tape the edges of the paper together over the object. Place the paper and object on a plastic or aluminum plate. Hold a magnet under the plate and move it back and forth. The paper will move!

How it works: The iron-containing metal object concealed in the paper is attracted to magnets. When you move a magnet under a nonmagnetic object such as a plastic or aluminum plate, magnetic force passes right through the plate, attracting the iron-containing object.

More science fun: Test different metals to see which ones are attracted to the magnet when they are concealed in the paper. Try different sizes and shapes of iron-containing metal objects (such as bolts, nuts, and nails) to see which are most strongly attracted by the magnet.

Flying Things

FLYING SPINNING PLATE

You will need: Paper, plastic, or polystyrene foam plate, 8 pennies or other coins, tape, scissors

What to do: Cut all but ¼ inch (.5 cm) of the curved edge from a paper, plastic, or polystyrene foam plate. Discard the outer rim. Tape 8 pennies at equal distances around the edge of the plate. Go outdoors and throw the plate outdoors, with a spinning motion, just as you would toss a commercially made flying disk, such as a Frisbee®.

How it works: When the plate is tossed, the coins around its edge make it heavier than it was without coins. As it flies through the air, the weighted plate

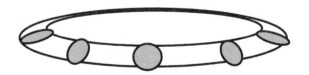

spins for a longer period of time and has more stability than an unweighted plate. This stability is similar to that of a rotating bicycle wheel.

More science fun: Try different sizes and types of plates. See how many and what type and weight of coins give the longest-flying and most stable flying plates.

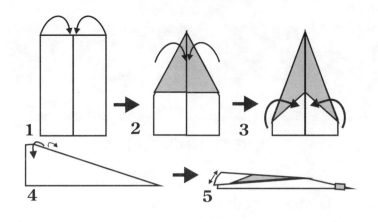

AIRPLANE

You will need: 8½ × 11 inch (22 ×28 cm) notebook paper, tape

What to do: Fold the paper in half down the long dimension (fig. 1). Fold the upper corners in so they meet at the center fold (fig. 2). Fold the outer edges of the parts you just turned in and part of the outer sides toward the center again (fig. 3). Fold the plane in half along the center crease (fig. 4) with the side folds on the outside. Fold down the sides of the wings 1 inch (2.5 cm) out from the center fold (fig. 5). Tape the nose of the plane (to keep the sides together and to add weight to the front of the plane). Fly your airplane!

How it works: The airplane flies because it is lightweight, has a streamlined shape which cuts easily through air, and has wings which help to slow it as gravity pulls it to earth.

More science fun: Make an airplane of very lightweight but still relatively stiff paper, such as onionskin paper. Does it fly better than one made with notebook paper?

INDEX CARD HELICOPTER

You will need: Index card, jumbo paper clip, scissors

What to do: Cut small rectangles out of two of the four corners of an index card (see fig. 2). Cut the card partway down the other long edge between halves A and B (fig. 3). Attach a paper clip to the bottom tab of the card. Bend the two blades A and B in opposite directions (fig. 4). Your helicopter is now ready to fly. Drop the helicopter from the top of some stairs, and watch it spin down, rotating all the way to the floor!

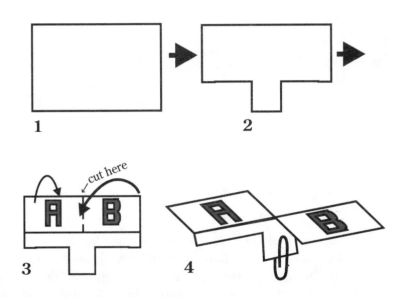

How it works: When air strikes blades A and B, air flows past the blades in opposite directions, giving a spinning motion to the helicopter. The weight of the paper clip keeps the bottom of the helicopter pointed toward the ground, holding the blades in position. As long as air keeps passing through the blades, the helicopter will to continue to spin.

More science fun: Try different sizes and numbers of paper clips. Try different sizes of index cards. Try substituting less stiff paper in place of the index card.

INDEX CARD–STRAW HELICOPTER

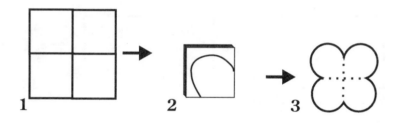

You will need: Index card, ballpoint pen, straw, scissors, masking tape, 2 paper clips

What to do: Cut the largest square you can from an index card. Fold the square in quarters. Draw a blade on one of the four quarters of the card as shown (fig 2). Keep the card folded in quarters. With scissors, cut through the four layers of the card where the blade is marked, making sure to keep all four blades connected at the center. Open out the shape. With the tip of a ballpoint pen, push a hole through the center of the card. Stick a straw ½ inch (1 cm) into the hole (fig. 4); widen

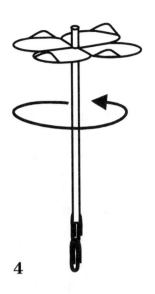

the hole as necessary. Attach two paper clips to the other end of the straw. Bend each of the edges of the helicopter blades a little, all in the same direction. Drop the helicopter from at least 5 feet (1.5 m) above the ground.

How it works: Since the edges of the helicopter blades are all bent in the same direction, when air strikes the helicopter blades, the helicopter rotates as it falls through the air.

4

More science fun: Try different sizes and numbers of paper clip weights. Try different sizes of index cards. Try substituting less stiff paper in place of the index card. What happens to the direction the helicopter spins in when you change the pitch (blade angle) of the blades to the opposite direction? Does the amount at which the blades are bent affect the speed of rotation of the helicopter?

PAPER PLATE HELICOPTER

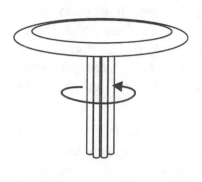

You will need: Paper, plastic, or polystyrene-foam plate, 4 flexible straws, tape

What to do: Tape the long parts of 4 flexible straws together to make a straw column. Bend the short ends of the straws at right angles to the straw column. Keeping the short straw ends at equal distances apart and at right angles to each other, tape them to the top surface of a paper, plastic, or polystyrene foam plate. To fly the helicopter, turn it so the straws are underneath, and hold the straw column between both palms of your hands. Quickly move your hands in opposite directions, keeping pressure on the straw column. When you have moved your hands as far as you can, pull them away from the straw column. The helicopter will fly, spinning through the air!

How it works: Rotating the straw column also rotates the attached plate. Since the plate has a greater diameter and mass than the straw column, this gives the helicopter significant spinning speed and stability. Think about the relationship of bicycle foot pedals to the rear tire. Depending upon the gear ratio, just one turn of the foot pedals can cause the much larger rear tire to rotate several times. This increases the speed at which the bicycle can go.

More science fun: Try using more or fewer straws. How many straws give the best helicopter? What kind and size of plate is best?

PLATE KITE

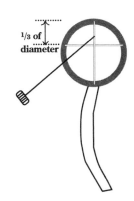

1/3 of diameter

You will need: Paper or polystyrene foam plate, 4 plastic drinking straws, spool of strong thread or twine, crepe paper, plastic wrap, scissors, tape

What to do: Cut the center out of a plate, leaving a ½ inch (1 cm) rim. Discard the center and save the rim. Stick the end of one straw into the end of another straw. Tape them together. Repeat with 2 other straws. Tape the ends of one of the double-length straws (held horizontally) to the upper third of the under-surface of the plate rim. Tape the second double-length straw at right angles across the first straws on the diameter of the plate rim (see drawing). Cut off any parts of the straws that stick out past the edge of the rim. Tape plastic wrap to the rim, stretching it as you do so. Cut off any plastic sticking out past the edge of the rim. Tape a 3-foot (1 m) long piece of crepe paper to the bottom of the kite. Tape the loose end of a spool of thread or twine to the vertical center of the kite on the plastic wrap surface (not on the straw), one-third from the top. Fly your kite! (It will fly nicely in a gentle breeze.)

How it works: When air passes over and around the front part of a kite, an area of lower pressure is created on the back side. Lifted by the air pressure difference between its front and back sides, the kite rises.

More science fun: Make kites from different sizes of plates. Which fly better: paper plate or polystyrene foam plate kites? Experiment with where the thread is attached to the kite; depending on the wind, the best position for the thread may not be exactly one-third of the distance from the top of the kite.

53

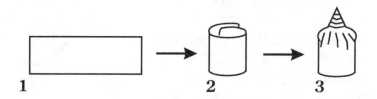

1   2   3

FINGERTIP PALM ROCKET

You will need: Lightweight paper

What to do: Cut a lightweight piece of paper 1¼ × 3 inches (3 × 7.5 cm). Wrap the strip around the end of your pointer into a curl (fig. 2). Twist about ¼ inch (2.5 cm) of the edge at the end of your pointer together into a tight cone (fig. 3). Make a circle with the thumb and pointer of your left hand. Put the rocket with twisted end up into the circle of your fingers so that it fits not too loosely or tightly (you may need to experiment to see which is the best way to hold it). Cup the other hand and strike the fingers around the rocket from underneath up. The rocket will shoot up into the air!

How it works: Air from your cupped hand forces the rocket up. The better (tighter) the seal between the rocket and the inside of your hand, the farther the rocket will travel. Be sure not to squeeze the rocket too tightly, or it will not leave your hand.

More science fun: Experiment with different sizes and weights of palm rockets. Compare them with rockets made from other materials, such as chewing gum wrappers or aluminum foil.

Heat Experiments

SPINNING SPIRAL SNAKE

You will need: Paper, toaster, thread, scissors, tape

What to do: Cut a piece of paper into a spiral-shaped snake. Tape a piece of thread to the snake's tail and hold it over a toaster. The snake will slowly turn!

How it works: Hot air rising from the toaster pushes on the edges of the snake's spiral-shaped body. The snake rotates because the direction of the air rising through the snake is changed from rising straight up to rising in a spiral. The snake turns in the direction of the air flow.

More science fun:
Create different sizes and shapes of snakes to see which will rotate the best.

PAPER INSULATES WATER GLASS

You will need: Brown grocery paper bag or other heavyweight paper, 2 drinking glasses, ice water, hot water

What to do: Wrap one of two drinking glasses with heavyweight paper. Fill both glasses with ice water. In which glass does the ice melt faster? Repeat this experiment with hot water. Which of the two glasses cools off (loses heat) most quickly?

How it works: Paper insulates the drinking glass. An insulator keeps heat from being gained (by the glass of ice water) or lost (by the glass of hot water). Exposed to the outside air, the uninsulated glass comes to the same temperature as its surroundings more quickly than the insulated glass does. Thermos™ vacuum bottles conserve heat and cold in the same way, and have a vacuum (a place with no air) between their glass or plastic insulating walls. A vacuum is an outstanding insulator, since it does not conduct heat.

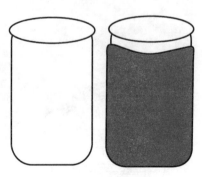

More science fun: Try different weights and types of paper. Which types make the best insulators? Does wet paper insulate as well as dry paper does? Do an experiment to find out.

BLACK AND WHITE PAPER
ON A SUNNY DAY

You will need: 2 drinking glasses or jars, black paper, white paper, tape

What to do: Wrap two drinking glasses or jars with paper—one glass with black paper, the other with white paper. Place both jars outside on a bright, sunny day. Check the temperatures of the jars after they have been standing ½ to 1 hour. Which is hotter?

How it works: The jar wrapped with black paper heats up more than the jar wrapped with white paper. This is because black paper absorbs, and white paper reflects, the sun's rays. When the rays are absorbed by a dark material such as black paper, they are converted from radiant energy to heat energy.

More science fun: Wrap glasses with other materials, such as aluminum foil or plastic. Do any of the glasses heat as much as they did with the black paper? Do they all heat up more than the glass wrapped with white paper? Note in particular how hot or cool the foil-wrapped glass is; foil reflects the sun's rays.

Light Experiments

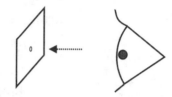

PINHOLE VISION-IMPROVER

You will need: Index card or paper, thumbtack or pin, book

What to do: With a thumbtack or pin, push a small hole in an index card or paper. Look at the small letters in a book through the paper pinhole. If you are nearsighted and wear glasses, take off your glasses. You will see both near and faraway objects more clearly through the pinhole than you do if you look at them without the aid of the pinhole.

How it works: When light rays reflected from the book pass through the tiny hole in the index card, the rays are pushed more closely together. The light rays become focused by a small part of the eye's lens on the retina (the light-sensitive back of the eye). This focusing effect causes blurred images to become clearer.

More science fun: Change the size of the pinhole. How large can the hole be before it no longer clarifies images?

TOP CHANGES STRAIGHT
TO CIRCULAR LINES

You will need: Index card, black marker, toothpick, ruler, masking tape, compass

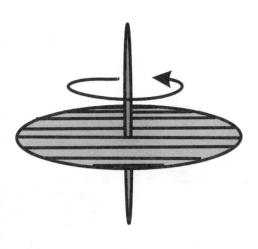

What to do: Draw eight parallel lines on an index card with a black marker. Cut a 2½ inch (6 cm) circle from the card. Push a toothpick through the center of the circle and tape the toothpick in place. With your thumb and forefinger, spin the toothpick shaft quickly. The card's straight lines will seem to change into circular lines! Also, you may see spots of color in the white spaces. If the disk is lit by a fluorescent light as it slows down, the disk's lines will seem to first go in one direction, briefly stand still, and then reverse direction.

How it works: This optical illusion is caused by the brain's being fooled by the spinning disk and its interaction with light. The eye interprets the eight straight lines as four concentric circles, with rings of dark lines separated by white spaces. The spots of color that may be seen in the white bands are caused by the brain's subtracting out one color more than others as black areas sweep over the white areas. The reason the lines change direction under fluorescent light as the disk slows down is that fluorescent light acts like a high-speed strobe light. Unlike an incandescent light, which glows continuously, fluorescent lights blink on and off at a speed that is normally too fast for the human eye to detect. But if spun fast enough, the disk will show the high-speed blinking of the fluorescent light.

More science fun: Make different sizes of disks. How many disks can you spin at the same time? Do they all show the same optical illusions?

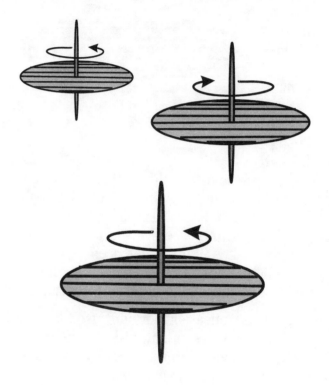

TOP CHANGES BLACK CURVED LINES TO COLORS

You will need: Index card, black marker, toothpick, ruler, masking tape, compass

What to do: Draw and cut out a 2½ inch (6 cm) circle from an index card. With a black marker, draw the pattern shown on the circle. Push a toothpick through the center of the circle and tape it in place. Spin the toothpick shaft quickly. You will see colors appear! When you reverse the direction of spin, the colors change!

How it works: This optical illusion is caused by the brain's being fooled into thinking that the alternating regions of black and white are instead colored. White is all colors mixed together. Black is the absence of colors. When the eye sees a blurry combination of white and black, it interprets it as a color. The nature of the color depends upon how much black is "mixed" with the white region, and how fast (and in what direction) the disk is spinning.

More science fun: Draw different patterns on the disks. Do you see other optical illusions when you spin the disks? Do you see more colors? Fewer colors? Other colors? How intense (bright) are the colors?

BLACK-AND-WHITE TO COLOR
SPINNING DISK

You will need: large, 5 × 8 inch (12.5 × 20 cm) index card, 8 feet (2.5 m) string, scissors, tape, ballpoint pen, black marker, toothpick, compass

What to do: With a compass, draw a 5 inch (13 cm) circle on an index card and cut it out. Draw black patterns on the circle as shown. With a ballpoint pen, punch two holes, 1 inch (2.5 cm) apart from each other, and each ½ inch (1 cm) from the center of the circle. Push part of the string through one hole, then through the other hole. Tie the ends of the string together. Adjust the string so that the disk is at right angles to the string. Tape a toothpick to the string and to one side of the disk. Grasp the ends of the string, with one loop in each hand. Move both hands together in a circular motion, causing the disk to rotate. The string will wind up tightly. Now pull your hands apart quickly. The disk will spin rapidly, causing colors to appear (look quickly, before the disk slows down!)

How it works: When the wheel is spun, the alternating

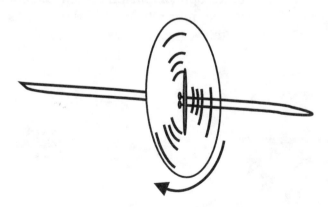

regions of black and white become a blur to the eye. Since the brain cannot distinguish between the black and white regions, it attempts to compensate by seeing them as a color. The nature of the color depends upon how much black is "mixed" with the white region, and how fast (and in what direction) the disk is spinning.

More science fun: Draw different patterns on other disks. Which patterns give the most easily seen colors when the wheels are spun?

COLOR-MIXING WHEEL

You will need: Large, 5 × 8 inch (12.5 × 20 cm) index card, 8 feet (2.5 m) of string, colored markers, ballpoint pen, scissors, tape, toothpick, compass

What to do: Using a compass, draw a 5 inch (12.5 cm) circle on an index card and cut it out. Draw wedges of different colors from the center of the circle. With a ballpoint pen, punch two holes, 1 inch (2.5 cm) apart from

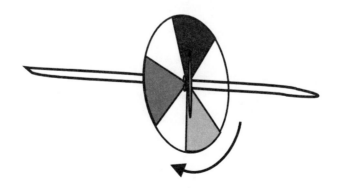

each other, and each ½ inch (1 cm) from the center of the circle. Push part of the string through one hole, then through the other hole. Tie the ends of the string together. Adjust the string so that the disk is at right angles to the string. Tape a toothpick to the string and to one side of the disk. Grasp the ends of the string, with one loop in each hand. Move both hands together in a circular motion, causing the disk to rotate. The string will wind up tightly. Now pull your hands apart quickly. The disk will spin rapidly. As the disk spins, its colors blend, causing the disk to appear gray or light black.

How it works: When the wheel is spinning rapidly, the human brain cannot distinguish the colors on the wheel, so it interprets them as a single color. If the disk has approximately equal-sized wedges of several colors, the disk will appear gray or gray-brown. If the disk contains mostly one color, then the disk when spun will display that color, but it will look duller (grayer) than the color of the disk at rest.

More science fun: Change the relative sizes of the wedges. How does this affect the spinning disk's color and the intensity (brightness) of the color?

Motion and Inertia Experiments

ROLLING STAR HOOP

You will need: 2 paper plates, scissors, tape, ballpoint pen

What to do: Tape the bottoms of two paper plates together, making sure that the outer edges are lined up. Punch a hole in the center with a ballpoint pen. Cut away a circle in the center of the plates, leaving a few inches before the rim. Cut 8 or more triangles through the remaining parts of both plates, but don't cut through the rims. Bend out all of the triangles away from the plate center towards the rim so that half of the triangles are pointed to one rim, half to the other (see picture). You now have a star hoop! Take it outside on a breezy day and stand it on its edge. It roll down the sidewalk or street when it catches the breeze!

How it works: The wind catches the ends of the triangular-shaped pieces sticking out from the hoop. The force of the wind against the pieces causes the hoop to roll.

More science fun: Change the size and number of the triangles on the Star Hoop. What size and number give the best (fastest moving) Star Hoop?

PAPER DROP TRICK

You will need: Thin strip of writing paper (or index card or dollar bill)

What to do: Hold a thin strip of writing paper (or index card or dollar bill) pointing downward between your thumb and forefinger. Ask a friend to place a thumb and forefinger about 1 inch (2.5 cm) below where you're holding the paper. Challenge your friend to catch the paper when you drop it. Drop the paper. She or he will not be able to catch it!

How it works: Human reflexes are too slow (about ½.. second) to catch the dropping piece of paper. By the time that the message that the paper has been released has been sent from your friend's eyes to his brain and then back from his brain to his hand and finger muscles it is too late—the paper has already dropped through his fingers!

More science fun: How far above your friend's fingers you can hold the piece of paper without his being able to catch the paper when it is dropped? Does the shape of or weight of the paper make a difference in how readily it can be caught?

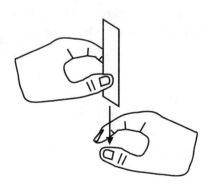

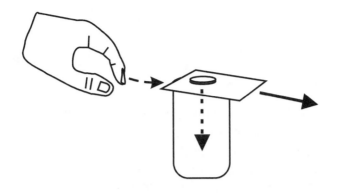

COIN DROP INTO GLASS

You will need: Index card, coin, drinking glass

What to do: Place a coin on top of an index card. Place the index card on the mouth of a drinking glass. With your thumb and forefinger, flick the index card quickly away from the glass. The coin will fall straight down into the glass!

How it works: The coin wants to stay in one place when the index card is flicked away from underneath it because of a property of matter called inertia. Inertia is the property that an object has to remain in its current location. Objects at rest tend to remain at rest. Similarly, objects in motion tend to remain in motion.

More science fun: Try different sizes of coins and index cards to see which sizes work the best. Note: if you do not flick the card fast enough, the coin will move off the glass along with the index card and will not fall into the glass. How quickly must you flick the index card in order for the coin to drop straight into the glass?

STATIONARY ERASER
ON PAPER STRIP

You will need: Rectangular pencil eraser, 1 × 5 inch (2.5 ×13 cm) strip of paper, ruler or pencil

What to do: Place a pencil eraser on one end of a 1 × 5 inch piece of paper. Position the eraser and paper so that 3 to 4 inches (7.5 to 10 cm) of the paper is hanging over the edge of a table. Grasp the hanging end with one hand. With a ruler or pencil, strike the paper strip abruptly close to the edge of the table. The paper will be pulled out from underneath the eraser, and the eraser will stay on the table. If you do this quickly enough, the eraser may even remain upright without falling over!

How it works: Objects at rest tend to remain at rest, just as objects in motion tend to remain in motion unless acted on by outside forces. The eraser is not pulled off the table along with the paper because the eraser's inertia keeps it in place. If you were to pull the paper too slowly, both eraser and paper would fall off the table.

More science fun: Try other objects in place of the pencil eraser, such as a chalkboard eraser, a heavy metal bolt, or a spoon. Does the weight of the object affect its ability to stay in one place?

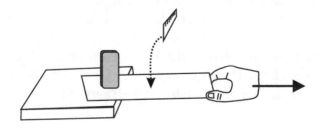

PROPELLER

You will need: Index card, pen, straw, scissors, masking tape, toothpick

What to do: Cut the largest square you can from the index card. Fold it in quarters (fig. 1). Draw a blade on one of the four quarters of the card (fig. 2). With scissors, cut through the four layers of the card where the blade is marked, making sure to keep all four blades connected at the center. Unfold the cut shape (fig. 3). Stick a toothpick halfway through the center of the shape. Bend each of the edges of the helicopter blades a little, all in the same direction. Place the bottom part of the toothpick into one end of a soda straw. Blow on the side of the index card blades. The toothpick (and blades) will rotate while resting in the straw.

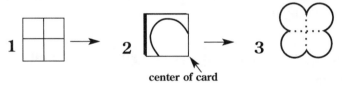

center of card

How it works: Since the blades of the propeller are all bent in the same direction, the propeller turns when you blow on it.

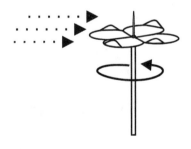

More science fun: Try different sizes of propellers. What happens to the direction the helicopter spins in when you change the pitch (blade angle) of the blades to the opposite direction? Does the amount which the blades are bent affect how easily the propeller turns?

PINWHEEL

You will need: Square piece of paper 4 to 6 inches (10 to 15 cm) on each edge, paper clip, soda straw, scissors

What to do: Cut a pinwheel with four joined blades from a 4 to 6 inch square piece of paper as you did for the propeller (page 69). Bend up a corner of each blade. Unbend one end of a paper clip. Push the end of the paper clip first through the center of the blades, then through the top of a soda straw (see drawing). Bend the end of the paper clip around again to keep it from falling out of the straw. Your pinwheel will easily rotate (spin) if you blow on it; take it outside on a windy day, or walk around with it.

How it works: The pinwheel rotates because the blades on the wheel are all pointed in the same direction. When air hits the blades, the pinwheel rotates.

More science fun: Change the angle of the blades. Try larger or smaller blades. Which size and shape works the best?

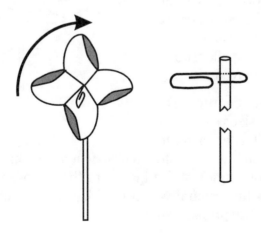

CARDBOARD TUBE BALL TWIRLER

You will need: Cardboard toilet paper tube, paper, string, scissors, masking tape

What to do: Crumple up two fist-sized paper balls, one larger than the other. Cut a piece of string 2 feet (60 cm) long. Pull the string through a toilet paper tube, with equal lengths sticking out of both ends of the tube. Tape one of the paper balls to one end of the string and the other paper ball to the other end. Hold the toilet paper tube upright, with the larger ball on the bottom and the smaller ball on the top. The heavier and larger ball will drop, pulling the lighter and smaller ball up against the top of the tube. Now twirl the tube around. The small ball jumps away from the top of the tube, in a circular path around the tube. As this happens, the large ball is pulled up against the bottom of the tube.

How it works: The light small ball surprisingly pulls up the heavy large ball when the tube is moved in a circle. This is because centripetal force of the small ball is greater than the gravitational force difference between the large ball and the small ball. The faster the ball and string are swinging around the toilet paper tube, the more quickly the large ball is pulled up.

More science fun: Change the size (and weight) of the balls. How light can the small ball be before it is no longer able to pull up the large ball?

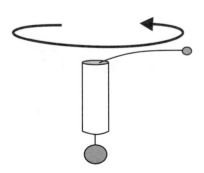

OATMEAL CARTON COME-BACK CAN

You will need: Oatmeal carton (including lid), ballpoint pen, large rubber band, heavy metal weight such as a bolt or nut, 2 nails or large paper clips

What to do: Punch holes in the centers of both ends of an oatmeal carton with a ballpoint pen or nail. Hang a heavy metal weight such as a bolt or nut on the center of the rubber band, inside the carton. Push an end of the rubber band through each end of the carton. Close the lid. Attach the ends of the rubber band to the outside ends of the carton by putting nails or paper clips through the rubber band loop and taping them in place. Roll the can along the floor. At the end of its roll away from you, it will roll back to you!

How it works: As the can rolls forward, it stores energy in the rubber band, which twists. At the end of the forward roll, the wound-up rubber band transfers energy back to the can, causing the can to roll back to you.

More science fun: Vary the weight of the object inside the can. Vary the size and length of the rubber band. Also try multiple rubber bands. Which returns to you more quickly: a can having just one rubber band or a can with three rubber bands?

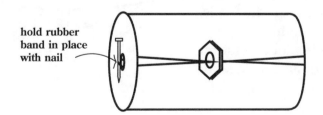

hold rubber
band in place
with nail

Noise- and Sound-Makers

PAPER CRUMPLING

You will need: Paper (notebook, construction, other types)

What to do: Slowly crumple a piece of notebook paper. Note the sound that it makes. Now crumple another piece of notebook paper, this time more quickly. The sound is louder than when it is crumpled slowly.

How it works: Sound is created when paper fibers rub against each other. The more brittle the paper and the more quickly it is crumpled, the louder the sound.

More science fun: Crumple different types of paper. Which types make the most noise?

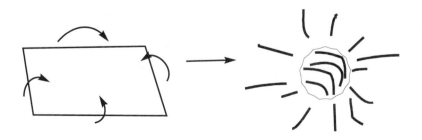

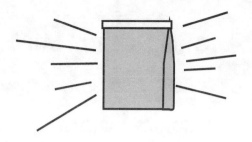

POPPING PAPER BAG

You will need: Brown paper bags (lunch size and large grocery size), masking or duct tape

What to do: Stick your hand inside a small (lunch size) brown paper bag. Shake the bag if needed to fully open it up. Tape the top shut. Lay the bag on the floor and quickly jump onto it. The bag will give a loud popping sound.

How it works: When the bag is stepped on and ruptures, the loud sound you hear is caused by the sudden release of air. The quicker the release of air, the louder the sound.

More science fun: Open up, tape, and pop a large brown paper grocery bag. Does the larger bag give a louder sound than the smaller lunch bag?

EXPLODING ORANGE JUICE CARTON

You will need: Small, ½ to 1 pint (.2 to .47 L) orange juice or milk carton

What to do: Close the top of an empty small orange juice or milk carton. Place the carton on a hard surface, such as a tile floor or a sidewalk. Quickly smash the carton with your foot. It will "explode" with a very loud noise.

How it works: The sound of the carton "exploding" is created by the sudden release of air. The quicker the release and the greater the amount of air which is released, the louder the sound. Juice and milk cartons will normally produce a louder noise than paper bags. This is because the cardboard of juice and milk cartons is stiffer (and less leaky) than the paper in paper bags.

More science fun: Try larger, 1 quart or ½ gallon (1 L or 3.8 L) juice or milk cartons. When stomped upon, do these sizes make a louder noise than the smaller cartons?

BUZZING PAPER SHEETS

You will need: 2 sheets of writing paper

What to do: Hold two sheets of writing paper against each other. Blow between them. You will hear a buzzing sound.

How it works: The paper sheets are drawn together because moving air between them has a lower pressure than the air on the outside of the sheets. The papers pull together and push apart several times per second. This repeated contact makes a buzzing sound.

More science fun: Try different types, weights, stiffnesses, and sizes of paper sheets. Which give the loudest sound?

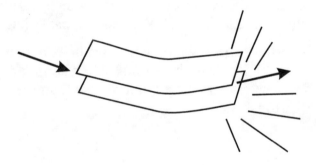

SCREECH WHISTLE

You will need: 1½ × 4 inch (4 × 10 cm) piece of paper, scissors

What to do: Bend the paper strip in half widthwise, giving a 2 inch long (5 cm) V-shaped piece. Bend up the ends of the V by ½ inch (1 cm). Cut a ½ inch (1 cm) slit in the center of the paper. Hold the ends of the paper strip to your lips and blow through the open (V-shaped) part of the paper strip. You will get a hideous screeching sound!

How it works: The paper whistle vibrates as air passes between the sheets of paper and out through the sides of the paper strip and the slit at the middle of the strip. Because the pressure of moving air is lower than the pressure of static (still) air, the opposite sides of the paper strip and edges of the slit in the paper are drawn together. The buzzing sound is caused by the repeated contact of the sides and edges of the paper and slits as they rapidly pull together and push apart.

More science fun: Experiment with different sizes and types of paper to see which give the best sounds.

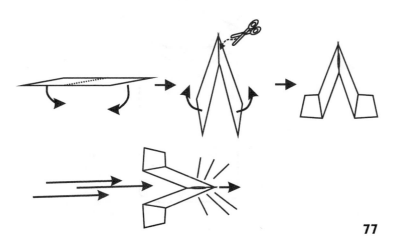

OATMEAL CARTON DRUM

You will need: Oatmeal carton, plastic wrap, large rubber band or duct tape, pencil or pen

What to do: Remove the lid from an oatmeal carton. Wrap the mouth of the carton with plastic wrap. Tightly secure the wrap in place with a large rubber band or with duct tape. Tap gently on the wrap with a pen or pencil to make drum beats.

How it works: When you tap on the plastic wrap, kinetic energy from the moving pen or pencil is converted into sound, another form of energy.

More science fun: Try stretching plastic wrap over the mouths of other containers, such as drinking glasses or empty peanut butter jars. Do they sound different from the oatmeal carton? Which give the loudest sounds?

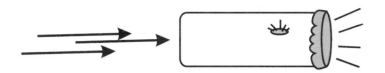

CARDBOARD TUBE KAZOO

You will need: Cardboard toilet paper tube, waxed paper, rubber band, scissors

What to do: Cut a small hole near the end of a toilet paper tube. Wrap waxed paper around the end of the tube. Fasten the paper in place with a rubber band. Blow into the other end of the tube. You will get a buzzing sound. By changing the pitch of your voice (humming), you can make music with your kazoo!

How it works: As air is blown into one end of the tube it escapes from the other (waxed paper) end. The waxed paper vibrates against the toilet paper tube, producing sound. The sound hole helps to regulate the air flow. The larger the hole, the less air that passes between the waxed paper and the end of the toilet paper tube.

More science fun: Try other types of paper, such as tissue paper or newspaper. Which type makes the best and loudest sounds?

FLAPPING WASHER SURPRISE

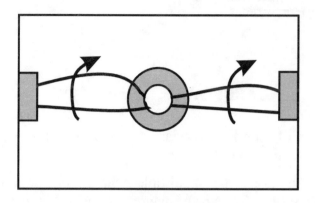

You will need: 3 × 5 inch (7.5 × 13 cm) index card, metal washer, 2 rubber bands, masking tape

What to do: Loop two rubber bands through a metal washer. Tape the free ends of the rubber bands to a 3 × 5 inch (7.5 × 13 cm) index card. Flip the washer over and over many times, until the rubber bands are tight. Then slip the index card into an envelope. Hand the envelope to a friend. When a friend opens the envelope, your Flapping Washer Surprise will jump around and make lots of noise!

How it works: When the index card is pulled out of the envelope, the washer and rubber bands, no longer held flat inside the envelope, are free to unwind. The loud rattling noise which you hear is created by the washer as it flaps rapidly and repeatedly against the index card.

More science fun: Hide the "Surprise" in other places, such as under a book, under a dinner plate, or under a sofa pillow. Watch how excited the next person will be who finds it!

PAPER POPPER

You will need: Lightweight writing paper size 8½ × 11 inches (22 × 28 cm)

What to do: Fold the piece of lightweight writing paper in half down the middle, along both the long (C to C) and short (B to B) dimensions (fig. 1). Unfold the paper. Fold up each of the four corners to meet the long axis (C to C), creating four triangles (fig. 2). Fold the paper in half along the long axis, with the triangles on the inside of the fold (fig. 3). Fold up the lower corner of the paper to the right (fig. 4).

Fold down the top corner (C) of the paper, to meet the

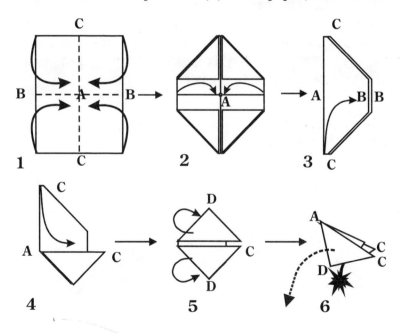

lower corner which you have just folded (fig. 4). Fold back each of the corners (D and D) to give the shape shown in fig. 6.

Holding corners C & C between your thumb and fore-finger, quickly jerk the paper popper downward. It will open up suddenly, giving a loud popping sound.

How it works: When paper abruptly strikes air, the paper's kinetic (motion) energy is converted into sound energy. The larger the paper, the louder the noise! After each use, refold the paper popper and crease the edges flat. This will help the popper to be reused several times (before it is eventually torn apart). If the popper does not pop on the first attempt, loosen up the "pocket" on the bottom of the popper by unfolding and refolding it loosely.

More science fun: Try other types and sizes of paper, such as newspapers and brown lunch or grocery bags (cut into flat sheets of paper). Heavy construction paper or large brown grocery bags work very well, holding their shapes and making a very loud noise!

PAPER CRACKER

You will need: Piece of lightweight writing paper, size 8½ × 11 inches (22 × 28 cm)

What to do: Fold a lightweight writing paper in 1 inch (2.5 cm) folds over and over on its short side until you have only 1 to 2 inches (3 to 5 cm) remaining unfolded. Fold back the paper cracker in the middle until its ends meet (A touching A and B touching B). Holding on to D with one hand, push corners A and A up through B and B until you have the final paper shape shown. Holding points B and B with your thumb and forefinger, jerk the paper cracker suddenly downwards. As air catches in the two pockets formed by the A-B pairs, the paper cracker will unfold very quickly, giving a loud cracking sound.

How it works: When paper abruptly strikes air, the paper's kinetic (motion) energy is converted into sound energy. The larger the paper, the louder the noise! After each use, refold the paper cracker and crease the edges

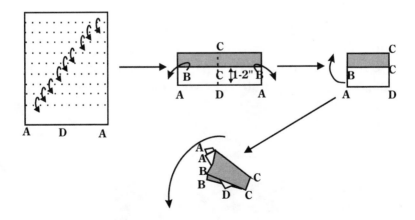

flat. This will help the cracker to be reused several times (before it is eventually torn apart).

More science fun: Try other types and sizes of paper, such as newspapers and brown lunch or grocery bags cut into flat sheets of paper. Large sheets of newspaper work very well, making a very loud noise!

COMB KAZOO

You will need: Hair comb (or pen, pencil, or finger), waxed paper

What to do: Fold a piece of waxed paper over a hair comb. Open your mouth slightly, place your lips on the waxed paper, and hum. The comb will buzz pleasantly.

How it works: As the waxed paper vibrates against your lips and against the comb, sound waves are formed. To create a melody, change the pitch of your voice as you blow on the comb.

More science fun: Instead of the comb, you can use any object, such as a pen or pencil, or the side of one of your fingers. If the waxed paper sticks to your lips, the paper can even vibrate after you remove the comb, pen, or pencil.

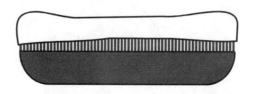

RUBBER BAND–PAPER PLATE HARP

You will need: Paper plate, 3 to 6 rubber bands, masking tape

What to do: Tape the ends of 3 to 6 rubber bands to a paper plate, bending the plate up slightly to create tension on the rubber bands. Pluck the rubber bands to give musical tones.

How it works: The rubber bands vibrate when plucked. These vibrations are transmitted through the air to your ear. The tighter the tension on the rubber band, the higher the musical note (or pitch).

More science fun: Try rubber bands having different lengths and thicknesses. How does this affect the tone which is produced? Are these tones higher or lower than the rubber bands you first attached to the paper plate?

PAPER SWINGER HUMMER

You will need: Three large, 5 × 8 inch (12.5 × 20 cm) index cards, string, rubber bands, tape, scissors, compass, ruler

What to do: Place three index cards on top of each other. Tape the edges together. Bend the stack of cards in thirds. Tape the ends together to make a triangular column. Cut a 1 inch (2.5 cm) diameter hole in each of the three sides of the column. Stretch and tape in place a rubber band over each of the three holes and also over the two open ends of the column. Tape a 2-foot (61 cm) piece of string to one triangular end of the index cards. Holding onto the loose end of the string, swing the hummer over your head quickly in a circle. It will produce a soft humming or whistling sound.

How it works: When air passes quickly enough over and through the rubber bands, the bands vibrate. This vibration is carried by air as sound waves. You hear it as a hum or a whistle.

More science fun: What happens to the sound when you speed up the swinging? Does it get louder? Does the pitch (tone) go up or down?

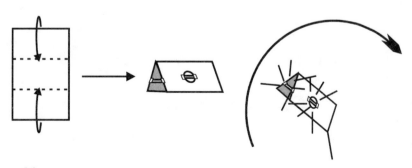

PAPER BULL-ROARER

You will need: Three large, 5 × 8 inch (12.5 × 20 cm) index cards, string, tape, scissors

What to do: Stack three index cards on top of each other. Tape the edges together. Bend the stack lengthwise to form a long, thin stiff strip. Tape one end of the cards to a 2 to 3 foot (60 to 90 cm) long piece of string. You've just made a Paper Bull-Roarer. To hear it, hold onto the unattached end of string and swing the Roarer around and around over your head.

How it works: When the edges of cards strike the air, they vibrate, creating the sound that you hear.

More science fun: Try using smaller index cards. Is the sound quieter? Make the bull-roarer larger by taping together additional index cards end-to-end. Is the sound of this larger Paper Bull-Roarer louder than the one you initially made?

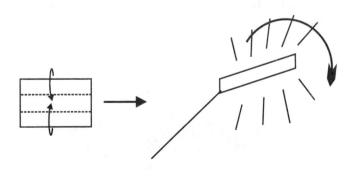

BIKE WHEEL WHAPPERS

You will need: Playing cards or index cards, duct tape

What to do: Tape playing cards, or index cards cut to the same size as playing cards, to the metal support forks of the front and back bicycle wheel. For the most secure attachment, use duct tape. Make sure that the cards are taped in a way that they are facing away from the direction of rotation of the bike wheel. When the wheel turns, the tip of the card will be drawn into the spokes. As the wheel turns further, the card tip is released. When the card attempts to go back to its original position, it strikes against the next spoke, creating sound energy. As the spokes rapidly strike the playing cards, a loud "twapping" sound results, which makes your bike sound like a small motorcycle!

How it works: Sound is created by the vibration of the cards when they are struck by the bicycle spokes. This vibration generates sound waves.

More science fun: Try different sizes and shapes of cards. Which ones give the best and loudest sounds? Try riding your bike at different speeds. Which speed gives the best sound?

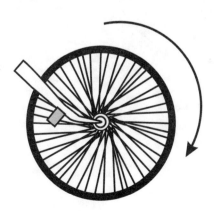

PAPER CUP TELEPHONE

You will need: 2 paper cups, ballpoint pen, 2 paper clips, 3 to 6 feet (90 to 180 cm) of string, duct tape, scissors

What to do: Punch holes into the bottom centers of two paper cups with a ballpoint pen. Push the ends of a 3 to 6 feet (90 to 180 cm) piece of string through the hole in each of the two cups from the outside to the inside, making sure that the open ends of the cups are facing away from each other. Inside each cup, tie the string end to a paper clip and tape the paper clip securely to each cup bottom with a strong tape, such as duct tape. With a friend, pull the cups away from each other until the string is tight. Talk into one of the cups. A person listening on the other end will hear you through the string. You now have your own private phone system!

How it works: When you speak into one of the cups, sound is carried along the string to the other cup. Sound is changed into string vibrations at the base of the first cup. These string vibrations are converted back into sound waves when they arrive at the second cup.

More science fun: Create a three- or four-party line by tying additional string and cup lines to the center of the two-cup line! Now you and up to three of your friends can have a party-line talk on your own personal telephone system!

PAPER BAG BUG AMPLIFIER

You will need: Brown paper bag, tape, bug

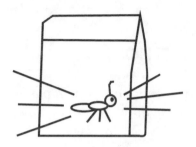

What to do: Place an insect such as a fly or beetle in a brown paper bag. Tape the top of the bag shut. Place the bag on its side and then lay your ear gently down on top of the bag. You will be able to hear the bug walking or buzzing around!

How it works: The surface of the bag acts to amplify (increase) the sound of the bug in your bag. This is because the surface area of the bag is much larger than the surface area of the bug's legs. Vibrations from the bug's legs are carried all over the paper in the bag, and are transmitted as sound waves.

More science fun: Place additional types of bugs into paper bags. Which make the most interesting sounds? Which are louder: flying or crawling bugs?

Note: Treat the bugs gently. When you finish your experiment, be sure to return them to wherever you found them.

Topology Experiments*

PAPER BALLS

You will need: 10 to 20 sheets of scrap paper

What to do: Place 10 to 20 sheets of scrap paper such as newspaper flat in a box. Notice how little space they occupy. Remove these sheets, crumple them up, return to the same box. The balls occupy much more space!

How it works: The balls have lots of air trapped inside them, and sheets of paper have very little air. Balls of paper are more compressible (squishable) than sheets of paper. These properties (lightness in weight and squishiness) make balls of paper such as newspaper an excellent choice for padding and protecting the contents of packages to be mailed.

More science fun: Which types of paper occupy the most space when wadded into balls ?

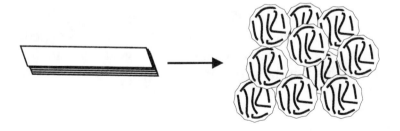

*Topology is a branch of mathematics that deals with the properties of geometric configurations which are unchanged by deformations such as stretching or twisting.

CONFETTI

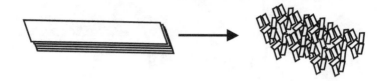

You will need: 10 to 20 sheets of scrap paper, scissors

What to do: Lay 10 to 20 sheets of a scrap paper such as newspaper flat in a box. Notice how little space they occupy. Remove these sheets and cut them into long thin strips. Cut these long strips into short pieces of paper. These little pieces of paper are called confetti. Place the confetti back in the same box. Notice that the confetti occupies more space than sheets of paper.

How it works: When paper is cut up into little pieces, the pieces have air trapped between them, more so than sheets of paper. For this reason, the confetti occupies more space than the uncut sheets of paper from which it came.

More science fun: Which types of paper occupy the most space when cut into confetti? When dropped, which types of confetti take the longest time to fall before they hit the ground?

PAPER BENDING

You will need: A piece of paper

What to do: Repeatedly fold a piece of paper in half, then in quarters, etc., again and again until you can fold it no more. How many times were you able to fold it? Were you surprised that you were not able to fold it more?

How it works: There is a limit to the number of times any piece of paper can be folded, no matter what its size; usually it is about 9 times. This is because the paper at the center bend gets too thick to tolerate any more bending. You'll notice that after the first couple of bends, the angle between the outer leaves of bent paper gets bigger and bigger (at first it's 0 degrees, and then it increases to close to 90 degrees, when it becomes impossible to bend the piece of paper any further).

More science fun: Try folding thinner and thicker pieces of paper. How many times can you fold them? Can you fold a large piece of paper more times than a smaller piece of paper?

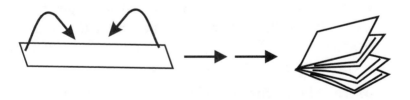

MAKE A SQUARE

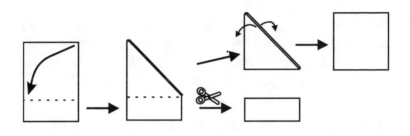

You will need: Paper, scissors

What to do: You can easily make a square from any rectangular piece of paper by folding over one corner at a 45 degree angle until the corner touches the opposite side. Cut off the excess paper. Unfold the folded triangular piece of paper. You have just made a perfect square!

How it works: A square is a rectangle whose four sides are of equal length. When you fold over the corner of the rectangle, the corner touches at a point on the long side of the rectangle which is equal in length to the short side of the rectangle. You've created a triangle. When you cut off the excess paper and unfold the triangle, you will get a square.

More science fun: How big or small can you make the square? Try different types and thicknesses of paper to see which most easily gives a perfect square.

FIND THE CENTER

You will need: Square piece of paper

What to do: You can locate the center of any square piece of paper by folding it in half, then in half again at right angles to the first fold. The corner where all four sections join is the center. Unfold the square and locate its center (where the creases cross each other).

How it works: A center is that point which is equidistant from all sides of a geometric figure. Since a square has four sides of equal length, when you fold it in half then in half again, you guarantee that all four sides are equidistant from a central point (where the creases cross each other at right angles).

More science fun: Try finding the center of pieces of paper which are not square. Start with rectangles and circles; then try odd-shaped pieces. For which ones can you find the center? For which ones can't you find the center?

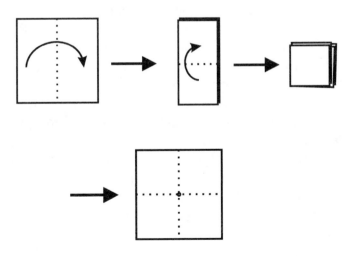

PAPER-STRETCHER COIN PUSH

You will need: Notebook paper, two coins of different sizes (such as a nickel and a quarter), pen or pencil, scissors

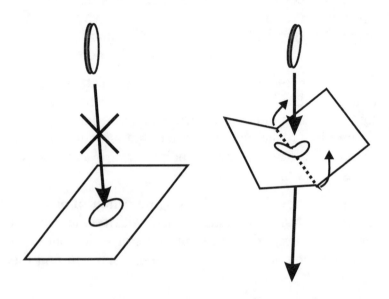

What to do: Select two coins of different sizes, such as a nickel (smaller coin) and a quarter (larger coin). Place the nickel on a sheet of notebook paper and draw a nickel-sized circle on the paper with a pen or pencil. Cut out the circle. Try to push a quarter through the nickel-sized hole without tearing the paper. You will not be able to. However, you can get the quarter through the hole in the paper by creasing the paper in half (across the center of the hole). The quarter can be pushed through the hole while you continue to hold the paper bent in half. To help get the coin through the hole, hold the opposite sides of the paper at the bend and gently pull up.

How it works: By bending the paper, the hole is made just large enough to accommodate the diameter of the quarter. Although the circle does not change in its overall circumference, by distorting the circle into an ellipse, in which one axis (distance across the ellipse through the center) is longer than the other axis, you provide enough room for the quarter to be pushed through. The long axis of the ellipse is wider than the diameter of the quarter. And since the quarter is thin compared to the shorter axis of the now elliptical hole, it can pass through.

More science fun: How small can the hole be before the quarter can no longer be pushed into it? How large can the size difference be between the smaller and larger coins? Try to push other small flat round objects through the hole, such as checkers. Will spherical objects such as marbles fit through the hole? Why or why not?

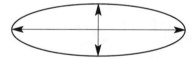

Ellipse, showing longest and shortest axis.

STRING AND BUTTON TRAP

You will need: 2 buttons or coins, string, paper, scissors, duct tape

What to do: Duct-tape a button or coin to each of the ends of a piece of string which is 1 foot (30 cm) long. Cut a hole that is slightly less than the diameter of the button or coin in a piece of paper. Try to get one button through the hole. You can't do it unless you crease the paper in half across the center of the hole. The button on one side of the hole can be pulled through the hole to meet the other button if you bend the paper in half while pulling on the string.

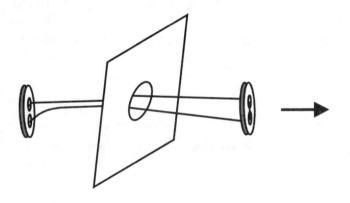

How it works: By bending the paper, the hole is made just large enough to accommodate the diameter of the button. Although the circle does not change in its overall circumference, by distorting the circle into an ellipse (in which one axis longer than the other axis), you provide enough room for the button to be pulled through. The long axis of the ellipse is wider than the diameter of the button. Since the button is thin relative to the shorter axis of the (now) elliptical hole, it can still squeeze through.

More science fun: How small can the hole be before the button or coin can no longer be pulled through it? Try to pull other small flat round objects, such as checkers, through the hole.

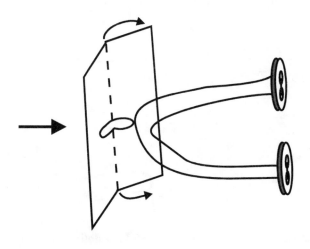

DOLLAR BILL PAPER CLIP TRICK

You will need: Dollar bill or other paper money or equally strong piece of paper, size about 3 × 6 inches (7.5 ×16 cm), 2 jumbo paper clips

What to do: Fold a dollar bill in thirds along its length into an S shape. Fasten one of the paper clips over one

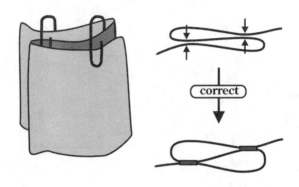

Looking down at the top edge of the folded dollar bill, you will see the S shape shown. Small arrows are positions of paper clips.

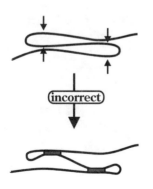

of the ends of the bill and the center fold (see drawing). Fasten the other paper clip over the other end in the same way. Each clip must pinch together two (and not three) thicknesses of the dollar bill. Make sure that the paper clips pinch the bill as shown in the "correct" drawing. Grasp one end of the bill with one hand and the other end of the bill with your other hand. Pull quickly. The paper clips will pop off the bill, linked together! Without your touching the paper clips, they have magically linked themselves together with a simple flick of your wrists!

How it works: By clipping the paper clips as described, when the dollar bill is pulled apart, one clip is forced to slide up into the loop of the other clip, linking it. To see exactly how this happens, pull the bill apart very slowly.

More science fun: Try other types and sizes of paper and paper clips. Which ones work the best?

INCREDIBLE EXPANDING
ZIG-ZAG LOOP

You will need: Piece of notebook-sized or larger paper, scissors

What to do: Fold a piece of notebook-sized paper in half along its long axis. Cut along the center of the fold, leaving ½ inch (1 cm) uncut at each end. Now cut thin slits into the folded and non-folded edges of the paper, alternating from one edge to the other. Cut both thicknesses at the same time, being careful to avoid cutting into the

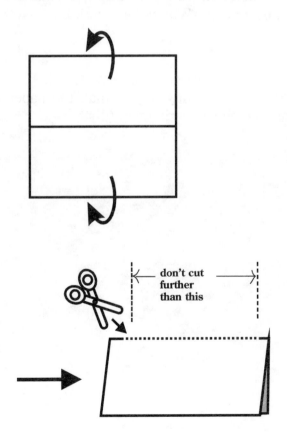

don't cut
further
than this

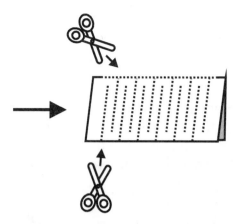

last ½ inch (cm) on each end of the paper. When you unfold the paper, you will find a loop that you (plus even a friend or two) can step into and pull up over your heads!

How it works: By cutting a long continuous winding path in the paper, you create a "hole" (paper loop) which is apparently bigger than the paper itself. The original paper perimeter (distance around all four edges) did not allow you to cut an ordinary hole that would be large enough for you to fit through. However, the length of the paper loop is not restricted to the length of the original perimeter. By cutting slits back and forth in the folded paper, a new and much longer perimeter is created. The length of this perimeter is limited only to how narrow you can cut the slits. The narrower the slits, the longer the perimeter, and the larger the hole.

More science fun: Although the paper loop can theoretically be made as large as you wish, its size is limited by the strength of the paper and the degree of precision with which you can cut. The thinner the cuts, the more likely the paper will tear. How large can you make the loop without having it tear?

MÖBIUS STRIP

You will need: 4 sheets of notebook paper, tape, marker

What to do: Place four sheets of notebook paper end to end and tape them together. Twist one end of this long piece of paper and tape it to the other (untwisted) end (connect A corners to B corners; see drawing). You

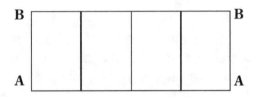

now have a closed loop with a single twist. This interesting loop is called a Möbius Strip. With a marker, draw a line down the center of the loop until you come back to where you started. You will find that the paper is marked on "both" sides of the loop! Since you have not lifted your marker from the paper, this proves this loop of paper has just *one* side! Now cut the loop down the center. How many loops did you get? How many were you expecting? Cut this longer loop down the center. How many loops did you get this time? How many were you expecting? Is there anything unusual about them?

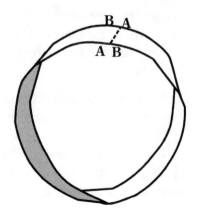

How it works: The topology (surface geometry) of an object such as a Möbius Strip is highly unusual. The loop gives you three surprises:

1. It has only one side, not two!
2. When cut down the middle the first time, it does not give the expected two loops, but rather one longer loop!
3. This longer loop, when cut down its center again, gives *two interlocked* loops!

More science fun: Cut each of the interlocked loops down the middle. Did you get the number of loops that you expected? Make another Möbius Strip. This time, do not cut it down the middle (½ to ½ ratio), but instead cut it into a ⅓ to ⅔ ratio. Any surprises? (You should get two interlocked loops, one of them twice as long as the other one!)

Water Experiments

PAPER DRINKING CUP

You will need: Large index card, 5 × 8 inches (12 × 20 cm), water

What to do: Fold an index card in half, then in half again at right angles to the first fold. Cut a quarter circle shape from the four stacked layers (see diagram). Open the cup by pulling out one of the outside layers from the other three layers.

How it works: The bottom of the pocket (the pocket formed when the outside layer is pulled away from the other layers) is the center (C) of the index card before it was folded. Water remains in the cup pocket until it

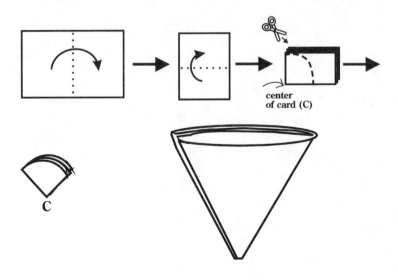

center
of card (C)

C

eventually soaks through the paper of the index card.

More science fun: Place waxed paper or aluminum foil on top of the index card before folding and cutting the card. These cups, lined with water-resistant waxed paper or waterproof aluminum foil, will hold water much longer than a plain paper index card will.

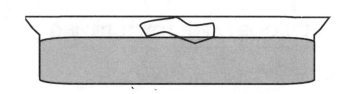

PAPER IN WATER

You will need: Paper (different types), water, dishpan

What to do: Place different kinds of paper in a dishpan of water. Which types fall apart most quickly? A good assortment of paper to try include: toilet paper, tissue paper, waxed paper, notebook paper, paper towels, and brown grocery bags.

How it works: The more absorbent and thinner the paper, the more quickly it falls apart in water. Paper is made by grinding up wood pulp with water, adding a few other ingredients, and then removing excess water. Therefore, by adding water to paper, you are reversing the paper-making process.

More science fun: Does paper fall apart faster in hot water than in room-temperature water? How quickly does it fall apart in ice-cold water?

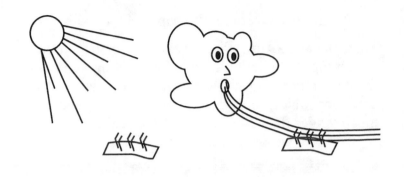

OUTDOOR PAPER TOWEL DRYING

You will need: Paper towel, water

What to do: Place a wet paper towel outdoors on a sunny day and also on a cloudy day. Does the towel dry faster than on the sunny day or on the cloudy day? Do the same experiment on breezy and calm days. Does the towel dry faster on a breezy day or on a calm day?

How it works: The paper towel dries out faster on a sunny day than on a cloudy day because radiant heat from the sun encourages the moisture in the paper towel to evaporate. Warm water has a higher vapor pressure than cold water, so warm water evaporates faster than cold water. The paper towel dries out faster on a breezy day than on a calm day because moving air carries off water faster than calm air. The air over a moist paper towel becomes saturated with water vapor on a calm day, but not on a breezy day.

More science fun: Try this same experiment on a cold wintry day, when ice and snow are on the ground. Do ice and snow disappear faster on sunny and breezy days than on cloudy and calm days?

WET PAPER TOWEL COOL TWIRL

You will need: Paper towel, string, waterproof tape such as duct tape, water

What to do: Tape a piece of string to a paper towel with a waterproof tape such as duct tape. Twirl the towel over your head, holding onto the string. Did the towel cool off? (It will not unless it is wet.) Now wet the paper towel and twirl it again. It will get cold.

How it works: Air passing across the surface of the moist paper towel evaporates the water absorbed in the towel. In order for water to evaporate, it must absorb heat from something. That "something" is the towel. As the water vapor gains heat in order to evaporate, the towel loses heat. A towel that loses heat becomes a cold towel.

More science fun: Does the towel cool off faster or become colder if it is twirled faster? Does it cool off faster on a hot day or on a cold day? Does it cool off faster after a rainstorm or just before a rainstorm?

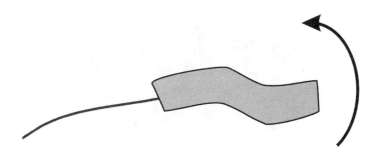

PAPIER-MÂCHÉ MOUNTAIN AND TREES

You will need: Paper towels, school glue, cornstarch, water, toothpicks, plastic bowl

What to do: Mix together cornstarch (4 tablespoons) and school glue (2 tablespoons) in a plastic bowl. Add just enough water (about 4 tablespoons) to make a thick paste. Tear a paper towel into 4 equal-sized pieces. Mix one of the pieces with the paste. Shape the paste into a small mountain. Stick toothpicks into the mountain; these toothpicks are the trees growing on your mountain. To speed drying, place the mountain and plastic bowl in a microwave oven and cook it for no more than 30 seconds at a time. Take the bowl out after each 30-second interval to see if the mountain has dried enough for the toothpicks to become glued to the mountain. Add treetops by sticking tiny globs of the pasty paper towel (called papier-mâché) to the tips of the toothpicks.

How it works: Glue and cornstarch bind paper fibers

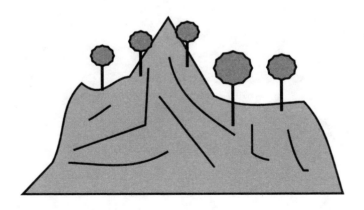

together. You can easily craft shapes from papier-mâché. When the shapes dry, they hold together very well, much better than shapes made from wet paper alone.

More science fun: Try different combinations of paper, glue, cornstarch, and water in a dishpan to see which gives the best (most easily shaped) paper structures. Which types of paper and glue work the best? Add some color to your creation as you make it by mixing brown food coloring (red + yellow + green) to the mountain and green food coloring (blue + yellow) to the treetops.

FILTER PAPER FUNNEL

You will need: Paper drinking cup (see page 106), paper towel, toothpick, wide-mouth jar or glass, orange juice or other mixture

What to do: Place a paper drinking cup over the mouth of a wide-mouth jar. Punch a hole in the bottom of the cup with a toothpick. Fold a paper towel the same way that the paper drinking cup was folded. Place the folded paper towel into the cup. You have made a filter funnel. The paper towel is the filter, and the paper drinking cup is the funnel. Pour a mixture such as orange juice into the filter funnel. The liquid will pass through the filter, and the solids will stay behind on the surface of the paper towel.

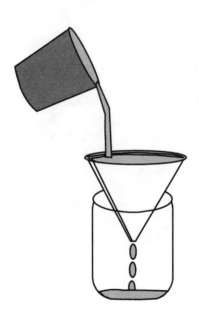

How it works: The fibers of the paper towel separate larger particles such as bits of orange from liquid. Particles smaller than the distance between the fibers of the towel will pass through; particles that are larger will not.

More science fun: Try other liquids that have particles in them, such as soups, juices, and hot chocolate with marshmallows.

BUBBLE SHEET

You will need: Waxed paper, paper towel, scissors, duct tape or other waterproof tape, dishwashing liquid, water, jar

What to do: Duct-tape a small piece of paper towel to the center of a sheet of waxed paper. Cut a hole in the center of both. Dip this hole into a bubble solution (1 part dishwashing liquid, 10 parts water). Wave the paper towel/waxed paper bubble sheet through the air. Lots of bubbles will come out of the hole!

How it works: Dishwashing liquid, when mixed with the correct ratio of water, forms nice bubbles. The absorbent edges of the paper towel soak up lots of bubble solution. As bubbles form, bubble solution is drawn to the edges of the hole, giving even more bubbles.

More science fun: Try different ratios of dishwashing liquid to water. Is 1 to 10 the best ratio? Try different brands of dishwashing liquid. Does the brand of dishwashing liquid change what the best ratio is?

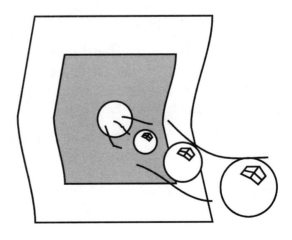

WATER DROPS ON PAPER

You will need: Paper (waxed and other types), water, toothpick, dishwashing liquid

What to do: Place three different kinds of paper next to each another: nonabsorbent (waxed), semi-absorbent (notebook paper, index cards, brown paper bags), and

absorbent (tissue and toilet paper, paper towels). Drop one or two drops of water on each one. Which type of paper has drops that are the roundest? What happens to the drop when you touch it with a toothpick moistened with dishwashing liquid?

How it works: Nonabsorbent paper, such as waxed paper, gives drops that are roundest. This is because water is repelled by (is pushed away by) wax. Semi-absorbent paper such as notebook paper is partially wetted. Since water is somewhat attracted to this type of paper, the drops are less rounded and spread out more. Absorbent paper (such as a paper towel) is attracted to water so much that a water drop cannot stay on top of the paper's surface at all. A split second after dropping onto the paper towel, the water drop completely disappears into the fibers of the towel.

More science fun: Dishwashing liquid, or any soap, lowers the surface tension of water. When soap touches water drops sitting on waxed paper, they become much less rounded, and they quickly spread out over the waxy surface of the paper. Which brands of dishwashing liquid cause water drops to spread out the fastest?

PAPER TOWEL AND TOILET PAPER
CAPILLARY PULL

You will need: Paper towel, toilet paper, water, 2 dinner plates, ruler

What to do: Dip a paper towel and a strip of toilet paper into two dinner plates containing water that is ¼ inch deep (.5 cm). Watch the water climb up the papers. Which of these two types of absorbent paper gives the fastest rate of water climb?

How it works: The finer and more extensive the fibers within the paper towel or toilet paper, the more effective the capillary action,* and the more quickly water will be drawn up into it.

More science fun: Try different brands of paper towels and toilet paper. Which brand gives the fastest rate of water climb? Which gives the slowest? The towel or toilet paper that gives the fastest water rise is the one which will probably soak up water the fastest. This is also a good test to measure the quality of brands of paper towels or toilet paper.

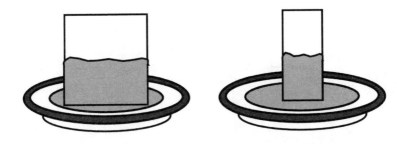

*Capillary action is the tendency of liquids in little tubes to rise in the tubes. Paper fibers act like little tubes and pull the water into the paper.

PAPER TOWEL CAPILLARY
SIPHON TUBE

You will need: 3 paper towels (connected and not torn), two glasses (tall and short), water, food coloring

What to do: Take the three connected (uncut) paper towels. Wrap them into a roll along their long side. Moisten the entire towel by dipping the towels into a tray of water. Stick one end of the towel roll into a tall glass that has been filled with water. Stick the other end into an empty short glass. Make sure that the end going into the short glass is lower than the end that is in the tall glass. In just a few moments, water will begin dripping into the short glass. This will continue as long as the water level in the tall glass is higher than that in the short glass, as long as the end of the towel stuck into the tall glass is still immersed in water.

How it works: Siphoning (water running up, then down, to an point lower than where it started from) takes place because of the capillary action of the towel,

where water flows through tiny channels in the paper fibers, seeking to reach as low a level as possible.

More science fun: Do an experiment to find out what gives better (faster) siphoning: a large vertical distance between the bottoms of the two glasses or a short vertical distance. As another experiment, siphon water containing one food color into a glass containing a different food color. Do you get a surprising new color?

COLORED WATER CAPILLARY PULL: COLOR CHANGE SURPRISE

You will need: White paper towel, dinner plate, water, food coloring, ruler

What to do: Sprinkle a few drops of blue food coloring into a dinner plate containing 1/8 inch (.3 cm) of water. Draw a line with yellow food coloring across a paper towel halfway up. Dip an edge of this paper towel that is parallel to the yellow band) into the water colored with blue food coloring. Hold up the towel as the blue food coloring rises up into the towel by capillary action. When the blue food coloring reaches the yellow band, the band will turn green!

How it works: Green color results when blue and yellow colors mix. Blue and yellow are two of the three primary colors (the other is red) which when mixed produce other colors.

More science fun: Try other types of color combinations: red + blue gives violet; red + yellow gives orange; red + yellow + green gives brown). What happens when you reverse the order of the colors in this experiment?

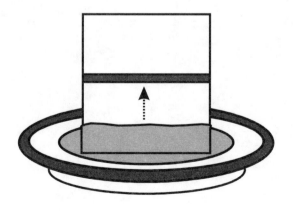

For example, does yellow food coloring in the dish and a blue band across the paper towel give a green color that is more intense than in the original experiment?

NEWSPAPER ZIG-ZAG STRIPS IN WATER

You will need: Newspaper, water, dinner plate or bowl

What to do: Cut up some ½ inch ×6 inch (1 cm × 15 cm) strips of newspaper. Bend them into zig-zag shapes. Place them in a dinner plate or bowl of water. In just a few seconds, the shapes will straighten out!

How it works: As water is drawn into the strips of newspaper by capillary action, the paper fibers swell. The swelling of the fibers causes the paper to straighten.

More science fun: Try other types of paper. Which type swells up the best? Is the performance of the paper related to its absorbency?

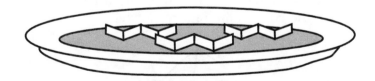

COFFEE FILTER CHROMATOGRAPHY

You will need: Coffee filter, water-based markers or highlighters, dinner plate, water

What to do: Draw a series of different colored lines with water-based markers about ½ inch (1 cm) from the edge of a coffee filter. Turn the filter upside-down and place it on a plate. Add a small amount of water to the plate, enough to touch all parts of the edge of the filter, but not touching the marker marks. As the water rises in the filter, the colored marks will move. And not only that, some of the colors will separate into several colors themselves!

How it works: Water rises up into the filter by capillary action. As it does so, it dissolves the water-based marker inks. Any inks which are composed of more than one color can be separated on the filter. Those colors that are hydrophilic (attracted to water) will move with the water front. Those colors that are less hydrophilic will move more slowly.

More science fun: Try different patterns and thicknesses of marks. Place some ice cubes around the base of the coffee filter just after the colors have begun to rise. Do the colors climb more slowly?

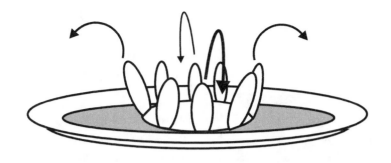

BLOSSOMING WATER FLOWER

You will need: Notebook paper, water, bowl, scissors

What to do: Cut a flower shape out of a piece of semi-absorbent paper such as notebook paper. Close the flower into a flower bud by bending the petals straight up and in towards the center of the flower. Float the flower on the surface of water in a bowl. As water is drawn up into the petals, the flower will open up.

How it works: When water flows by capillary action into the paper in the flower, the paper fibers swell. As the fibers swell, the petals start to move, "growing" upright.

More science fun: Try other types of paper. Which type of paper petals "grow" the most? Is the speed with which the petals open related to the absorbency of the paper?

SEED INCUBATOR

You will need: Paper towel, water, seeds, jar with lid

What to do: Wrap some seeds in a paper towel. Moisten the towel + seeds with water, put them into a jar, and tighten a lid on the jar. After a few days to a couple of weeks, the seeds will sprout.

How it works: Seeds require a constant and favorable moisture and temperature condition to sprout. By keeping the seeds wrapped in a moist paper towel and in a covered jar, you provide both of these conditions.

More science fun: Some seeds (such as radish) sprout very quickly; others (such as carrot) less quickly. Some (such as lettuce) require cool temperatures, and may need to be sprouted in your refrigerator. Others (such as melons) like warm temperatures and need a nice warm environment to sprout. Do some experiments to see which seeds sprout the fastest; which require the most (or least) water; and which sprout well in warm (or cool) temperatures.

Other Experiments

TEARING NEWSPAPER

You will need: Newspaper

What to do: Tear a piece of newspaper slowly from top to bottom. Then tear it the other way (side to side). You will get a reasonably smooth and straight tear in one direction and a jagged, irregular tear in the other direction.

How it works: Newspapers (and most other papers) have a "grain," a direction in which the wood fibers are lined up. If you tear parallel to these fibers ("with the grain"), the tear is straight. If you tear perpendicular to these fibers ("against the grain"), the tear is rough and jagged.

More science fun: Try tearing other types of paper. Which types tear differently in different directions?

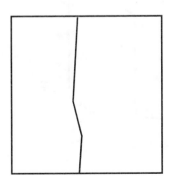 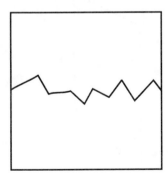

CUTTING A FRUIT WITH PAPER

You will need: A firm, sharp-edged piece of paper, such as high-quality printing paper or an index card; grape, peeled banana, or equally soft fruit or vegetable

What to do: With the edge of a sharp-edged piece of printer paper or index card, cut a grape, a peeled banana, or an equally soft fruit or vegetable. Experiment to see how many times or how deeply or you can cut it.

How it works: The edge of a good piece of printer paper is firm and does not absorb water as easily as cheaper paper does. The finer the edge and the less the water absorbency, the better the paper can cut. Index cards tend to cut better than thinner paper sheets because it takes longer for the fruit's water to soak into (and dull) the edge of the index card.

More science fun: Try other types of paper and other fruits and vegetables. Which work and which do not? Do cooked vegetables cut more easily than uncooked ones? Do canned fruits cut more easily than fresh fruits?

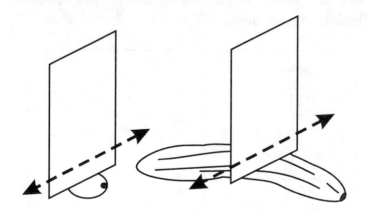

SAND FUNNEL

You will need: Paper, sand, salt, sugar, or rice (or other very small grainy material), strong tape such as duct tape

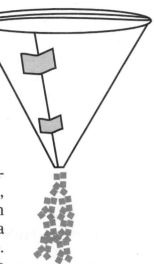

What to do: Form a piece of paper into a funnel; see instructions for making a paper drinking cup (p. 106) and making a filter paper funnel, (p. 112). Keep the funnel from unbending by fastening it with a strong tape such as duct tape. Punch a hole in the bottom of the funnel with a pencil. Pour sand (or similar grainy material such as salt, sugar, or rice) through it.

How it works: Gravity pulls the sand down into the funnel. The friction of the grains of sand against each other (and the larger grains of sand) slow down or even stop the flow. A number of factors enter into how fast the flow occurs. The steeper and smoother the sides of the funnel and the bigger the hole at the bottom, the more quickly small grainy material can flow through it. The wetter the material, the slower it flows. For example, moist salt or sugar will eventually cake up, causing big sticky lumps to form which flow very slowly, or not at all.

More science fun: Make funnels that have differing side angles and hole sizes. Make funnels of different materials (waxed paper, notebook paper, paper towels). Which ones work the best for sand? For salt? For sugar? For rice?

Index